Impossibility

*The Limits of Science and
the Science of Limits*

The meaning of the world is the separation of wish and fact.
KURT GÖDEL

Impossibility

*The Limits of Science
and the Science of Limits*

JOHN D. BARROW

UNIVERSITY OF CAMBRIDGE

OXFORD
UNIVERSITY PRESS

OXFORD
UNIVERSITY PRESS

Oxford   New York
Athens   Auckland   Bangkok   Bogotá   Buenos Aires   Calcutta
Cape Town   Chennai   Dar es Salaam   Delhi   Florence   Hong Kong   Istanbul
Karachi   Kuala Lumpur   Madrid   Melbourne   Mexico City   Mumbai
Nairobi   Paris   São Paolo   Singapore   Taipei   Tokyo   Toronto   Warsaw

and associated companies in
Berlin   Ibadan

© John D. Barrow, 1998

First published by Oxford University Press, Inc., 1998

First issued as an Oxford University Press paperback, 1999
198 Madison Avenue, New York, New York 10016

Oxford is a registered trademark of Oxford University Press

Library of Congress Cataloging in Publication Data
Barrow, John D., 1952–
Impossibility : The Limits of Science
and the Science of Limits / John D. Barrow.
1. Science–Philosophy.   2. Limit (Logic)   3. Gödel's theorem.
I. Title.
Q175.B2245   1998   501–dc21   97-35202
ISBN 0 19 851890 0
ISBN 0 19 513082 0 (pbk.)

10 9 8 7 6 5 4 3 2 1

Printed in the United States of America

In memory of
Roger Tayler

Preface

Both scientists and philosophers are much concerned with impossibilities. Scientists like to show that things widely held to be impossible are in fact entirely possible; philosophers, by contrast, are more inclined to demonstrate that things widely regarded as perfectly feasible are in fact impossible. Yet, paradoxically, science is only possible because some things are impossible.

The incontrovertible evidence that Nature is governed by reliable 'laws' allows us to separate the possible from the impossible. Only those cultures for whom there existed a belief that there was a distinction between the possible and the impossible provided natural breeding grounds for scientific progress. But 'impossibility' is not only about science. In the pages that follow we shall look at some of the ways in which the impossible in art, literature, politics, theology, and logic has stimulated the human mind to take unexpected steps: revealing how the concept of the impossible sheds new light on the nature and content of the actual.

The idea of the impossible rings alarm bells in the minds of many. To some, any suggestion that there might be limits to the scope of human understanding of the Universe or to scientific progress is a dangerous meme that undermines confidence in the scientific enterprise. Equally uncritical, are those who enthusiastically embrace any suggestion that science might be limited because they suspect the motives and fear the dangers of unbridled investigation of the unknown.

At the end of each century there seems to arise a stock-taking in science. We shall see that at the end of the last century the issue of the limits of science became a live one and attempts were made to pick out problems that could never be solved. These problems still make interesting reading. But what will people say about *our* concerns in a hundred years time? As we near the end of the twentieth century we look back on an extraordinary century of progress. Yet it is progress that possesses some extraordinary characteristics. A pattern has emerged in many spheres of inquiry in which a scientific theory becomes so successful in the quantity and quality of its accurate predictions that its practitioners start to wonder whether the end is in sight—whether their theory might be able to explain everything within its encompass. But then something strange happens. The theory predicts that it cannot predict. It turns out to be not simply limited in scope, but *self*-limiting. This pattern is so strikingly recurrent that it suggests to us that we can recognize mature scientific theories by their self-limiting character. Such limits arise not merely because theories are inadequate, inaccurate, or inappropriate: they tell us something profound about the nature of knowledge and the implications of investigating the Universe from within.

Our study of the limits of science and the science of limits will take us from the consideration of practical limits of cost, computability, and complexity to the restrictions imposed on what we can know by our location in the middle of the Nature's spectra of size, age, and complexity. We shall speculate about our possible technological futures and locate our current abilities on the spectrum of possibilities for the manipulation of Nature in the realms of the large, the small, and the complex. But practicalities are not the only limits we face. There may be limits imposed by the nature of our humanity. The human brain was not evolved with science in mind. Scientific investigation, like our artistic senses, are by-products of a mixed bag of attributes that survived preferentially because they were better adapted to survive in the environments they faced in the far distant past. Perhaps those ambiguous origins will compromise our quest for an understanding of the Universe? Next, we shall start to pick at the edges of possible knowledge. We shall learn that many of the great cosmological questions about the beginning, the end, and the structure of our Universe are unanswerable. Despite the confident exposition of the modern view of the Universe by astronomers, these expositions are invariably simplified in ways that disguise the reasons why we cannot know whether or not the Universe is finite or infinite, open or closed, of finite age or eternal. Finally, we delve into the mysteries of the famous theorems of Gödel concerning the limitations of mathematics. We know that there must exist statements of arithmetic whose truth we can never confirm or deny. What does this really mean? What is the fine print on this theorem? What are its implications for science? Does it mean that there are scientific questions that we can never answer? We shall see that the answers are unexpected and lead us to consider the possible meaning of inconsistency in Nature, of the paradoxes of time travel, the nature of freewill and the workings of the mind. Finally, we shall explore some of the strange implications of trying to pass from the consideration of individual choices to collective choices. Whether it is the outcome of an election or the making up of one's mind in the face of the brain's competing options, we find a deep impossibility that may have ramifications throughout the domain of complex systems.

Here, in this strange world of fundamental limits we learn that worlds that are complex enough for certain individualities to be manifest necessarily display an open-endedness that defies capture within the confines of a single logical system. Universes that are complex enough to give rise to consciousness impose limits on what can be known about them from within.

By the end of our journey, I hope the reader will have come to see that there is more to impossibility than first meets the eye. Its role in our understanding of things is far from negative. Indeed, I believe that we will gradually come to appreciate that the things that cannot be known, that cannot be done, and cannot be seen, define our Universe more clearly, more completely, and more sharply than those that can.

This book is dedicated to the memory of Roger Tayler, who sadly did not live to see it finished. His selfless service to his colleagues at Sussex and to the wider community of astronomers in Britain and around the world won him the respect, admiration, and friendship of scientists everywhere. He is greatly missed.

I would like to thank many people who helped me by their comments or advice, or who provided pictures and references, especially David Bailin, Per Bak, Margaret Boden, Michael Burt, Bernard Carr, John Casti, Greg Chaitin, John Conway, Norman Dombey, George Ellis, Mike Hardiman, Susan Harrison, Jim Hartle, Piet Hut, Janna Levin, Andrew Liddle, Andre Linde, Seth Lloyd, Harold Morowitz, David Pringle, Martin Rees, Nicholas Rescher, Mark Ridley, David Ruelle, John Maynard Smith, Lee Smolin, Debbie Sutcliffe, Karl Svozil, Frank Tipler, Joseph Traub, and Wes Williams. My wife Elizabeth helped in many practical ways, and accommodated innumerable new pieces of paper in the house with surprising good humour, whilst the subject of this book merely provoked our children, David, Roger, and Louise, to worry that there might indeed be fundamental limits on the use of the telephone.

Brighton
November 1997   J.D.B.

Contents

CHAPTER 1

The art of the impossible

If an elderly but distinguished scientist says that something is possible he is almost certainly right, but if he says that it is impossible he is very probably wrong.

ARTHUR C. CLARKE

The power of negative thinking

That's what I like about Lord Young. While you all bring me problems, he brings me solutions.

MARGARET THATCHER

Bookshelves are stuffed with volumes that expound the successes of the mind and the silicon chip. We expect science to tell us what can be done and what is to be done. Governments look to scientists to improve the quality of life and safeguard us from earlier 'improvements'. Futurologists see no limit to human inquiry, while social scientists see no end to the raft of problems it spawns. The contemplation by our media of science's future path is dominated by our expectations of great interventions: cracking the human genetic code, curing all our bodily ills, manipulating the very atoms of the material universe, and, ultimately, fabricating an intelligence that exceeds our own. Human progress looks more and more like a race to manipulate the world around us on all scales, great and small.

It would be easy to write such a scientific success story. But we have another tale to tell: one that tells not of the known but of the unknown; of things impossible; of limits and barriers which cannot be crossed. Perhaps this sounds a little perverse. Surely there is little enough to say about the unknown without dragging in the unknowable? But the impossible is a powerful and persistent notion. Unnoticed, its influence upon our history has been deep and wide; its place in our picture of what the Universe is like at its deepest levels is undeniable. But its positive role has escaped the critics' attention. Our goal is to uncover some of the limits of science: to see how our minds' awareness of the impossible gives us a new perspective on reality.

When we are young we think we know everything. But if we grow wiser as we grow older we will gradually discover that we know less than we thought. The poet W.H. Auden wrote of human development that

> between the ages of twenty and forty we are engaged in the process of discovering who we are, which involves learning the difference between accidental limitations which it is our duty to outgrow and the necessary limitations of our nature beyond which we cannot trespass with impunity.[1]

Our collective knowledge of the nuts and bolts of the Universe matures in a similar way. Some knowledge is simply the accumulation of more facts, broader theories, and better measurements by more powerful machines. Its rate of growth is always limited by costs and practicalities that we steadily overcome by attrition, little by little. But there is another form of knowledge. It is the awareness that there are limits to one's theories even when they are right. While the modest investigator might always suspect that there are things that will remain beyond our reach, this is not quite what we have in mind. There is a path of discovery that unveils limits that are an inevitable by-product of the knowing process. Discovering what they are is a vital part of understanding the Universe. This means that the investigation of the limits of our knowledge is more than a delineation of the boundaries of the territory that science can hope to discover. It becomes a crucial feature in our understanding of the nature of this collective activity of discovery that we call science: a paradoxical revelation that we can know what we cannot know. This is one of the most striking consequences of human consciousness.

There is an intriguing pattern to many areas of deep human inquiry. Observations of the world are made; patterns are discerned and described by mathematical formulae. The formulae predict more and more of what is seen, and our confidence in their explanatory and predictive power grows. Over a long period of time the formulae seem to be infallible: everything they predict is seen. Users of the magic formulae begin to argue that they will allow us to understand everything. The end of some branch of human inquiry seems to be in sight. Books start to be written, prizes begin to be awarded, and of the giving of popular expositions there is no end. But then something unexpected happens. It's not that the formulae are contradicted by Nature. It's not that something is seen which takes the formulae by surprise. Something much more unusual happens. The formulae fall victim of a form of civil war: they predict that there are things which they cannot predict, observations which cannot be made, statements whose truth they can neither affirm nor deny. The theory proves to be limited, not merely in its sphere of applicability, but to be *self-*limiting. Without ever revealing an internal inconsistency, or failing to account for something we have seen in the world, the theory produces a 'no-go' statement. We shall see that only unrealistically simple scientific theories avoid this fate. Logical descriptions of complex worlds contain within themselves the seeds of their own limitation. A world that was simple enough to be fully known would be too simple to contain conscious observers who might know it.

Of faces and games

I'm not young enough to know everything.

J.M. BARRIE[2]

Complete knowledge is a tempting pie in the sky. Although it appears in some commentator's minds as the obvious goal of science, it is a concept largely unknown within the writings of contemporary science. It is the hallmark of many varieties of pseudo-science, just as it pervades countless ancient myths and legends about the origin and nature of the world. These stories leave nothing out: they have an answer for everything. They aim to banish the insecurity of ignorance and provide a complete interlinked picture of the world in which human beings play a meaningful role. They remove the worrying idea of the unknown. If you are at the mercy of the wind and the rain it helps to personify those unpredictable elements as the character traits of a storm god. Even today, many spurious attempts to explain the world around us still bear this hallmark. Horoscopes seek to create a spurious determinism that links our personalities to the orientations of the stars. Uncertainties about tomorrow can be hidden behind vague generalities about the future course of events. It is strange how many inhabitants of modern democracies feel no qualms about living under an astral dictatorship that would plan their every thought and action.

This desire for complete seamless explanation infests most examples of crank science. When somebody mails me their explanation of the architecture of the Universe derived from the geometry of the Great Pyramid, or the cipher of the Kabbalah, it will usually display a number of features: it will be entirely a work of explanation; there will be no predictions, no tests of its correctness; and nothing lies beyond its encompass. It is not the beginning of any research programme. Beyond refutation, it is always the last word.

This desire to link all things together is a deep human inclination. It is not a modern fashion that arrived with the word processor. Its most famous ancient manifestation is to be found in the work of the ancient Pythagorean sect who mingled mathematics with mysticism.[3] They thought that number was a unifying principle in the Universe, so that anything that could be numbered was ultimately linked to other things with the same number. Numbers had meanings apart from their relationships with other numbers. Thus, musical harmony was linked to the motions of the heavenly bodies. The discovery that there were numbers that could not be represented by fractions precipitated a crisis so deep that these numbers had to be called 'irrational'. They appeared to lie beyond the complete arithmetic pattern of the Universe that the Pythagoreans had embroidered.

This unifying inclination of ours is a by-product of an important aspect of our intelligence. Indeed, it is one of the defining characteristics of our level of self-reflective intelligence. It allows us to organize knowledge into categories: to know vast numbers of thing by knowing rules and laws which apply in an infinite number of circumstances. We do not need to remember what the sum of every possible pair of numbers is: we need know only the principle of addition. The ability to seek and find common factors behind superficially dissimilar things is a prerequisite for memory and for learning from experience (rather than merely *by* experience). Some cultures have grown content with religious views of the world which are far less unified than others and have gods for every facet of life and Nature. In this sense, monotheistic faiths offer the most economical theological conception: by contrast, faiths with many disparate deities vying for influence seem less appealing.

All human experience is associated with some form of editing of the full account of reality ('we cannot bear too much reality'). Our senses prune the amount of information on offer. Our eyes are sensitive to a very narrow range of frequencies of light, our ears to a particular domain of sound levels and frequencies. If we gathered every last quantum of information about the world that impinged upon our senses they would be overwhelmed. Scarce genetic resources would be lopsidedly concentrated in information-gatherers at the expense of organs which could exploit a smaller quantity of information in order to escape from predators or to prey on sources of food. Complete environmental information would be like having a one-to-one scale map.[4] For a map to be useful it must encapsulate and summarize the most important aspects of the terrain: it must compress information into abbreviated forms. Brains must be able to perform these abbreviations. This also requires an environment that is simple enough and displays enough order, to make this encapsulation possible over some dimensions of time and space.

Our minds do not merely gather information; they edit it and seek particular types of correlation. They have become efficient at extracting patterns in collections of information. When a pattern is recognized it enables the whole picture to be replaced by a briefer summary form which can be retrieved when required. These inclinations are helpful to us and expand our mental powers. We can retrieve the partial picture at other times and in different circumstances, imagine variations to it, extrapolate it, or just forget it. Often, great scientific achievements will be examples of one extraordinary individual's ability to reduce a complex mass of information to a single pattern. Nor does this inclination to abbreviate stop at the door of the laboratory. Beyond the scientific realm we might understand our penchant for religious and mystical explanations of experience as another application of this faculty for editing reality down to a few simple principles which make it seem under our control. All this gives rise to dichotomies. Our greatest scientific achievements spring from the most insightful and elegant reductions of the superficial complexities of Nature to reveal their underlying simplicities, while our greatest blunders often arise from the oversimplification of aspects of reality that subsequently prove to be far more complex than we realized.

Our penchant for completeness is closely associated with our liking for symmetry. We have a natural sensitivity for pattern and an appreciation of symmetry that quickly picks up subtle deviations from perfect symmetry. Our desire for a full and perfect description of the world owes much to this curious sensitivity. Where does it originate?

A powerful means of understanding why we possess many odd abilities is to recognize that our mental faculties evolved several million years ago in environments that were very different from those in which we now live. In that primitive environment certain sensitivities would tend to enhance the survival prospects of those that possessed them with respect to those who did not. Those attributes which made survival more probable would be the expression of some complex genetic cocktail with no predetermined purpose. Although one feature of an attribute might aid survival, there might be by-products of this attribute which showed up subsequently in all sorts of unexpected ways. Many of our aesthetic sensitivities have arisen in this indirect manner. Accordingly, we can identify good evolutionary reasons why we might be expected to have developed an acute appreciation for symmetry. If we look at the natural environment we see that lateral (left–right) symmetry is a very effective discriminator between living and non-living things in a crowded scene. You can tell when a living creature is looking at you. This sensitivity has a clear survival value. It enables you to recognize potential predators, mates, and meals. This biological source of our appreciation of symmetry is supported by the fact that our most acute sensitivity for symmetry is manifested in our appreciation of the human form, especially the face (Fig. 1). Symmetry of bodily form—especially that of the face—is our most common initial indicator of human beauty, and we go to enormous lengths to enhance it and protect it.[5] In lower animals it is an important indicator of mates. In humans it has had all manner of by-products which influence our aesthetic appreciation and underlie our acute sensitivity to patterns, symmetry, and form. Remarkably, no computer has yet managed to reproduce our many levels of visual sensitivity to patterns.[7]

This sensitivity means that deviations from symmetry are quickly identified and have a sophisticated interpretation all their own. Because they capture our attention so dramatically they are much used in (English) humour. Try the

Fig. 1.1 An average human face, displaying lateral symmetry.[6]

effect of the following classic deviation from the traditional anapaestic symmetry of the limerick form:

> There was a young man of Milan
> Whose rhymes they never would scan;
> When asked why it was,
> He said, 'It's because
> I always try to cram as many words into the last line as ever I possibly can.'

A microcosm of our attitudes towards completeness can be found in the world of games. Simple games, like noughts and crosses, are entirely predictable. With a little thought you can devise a strategy that prevents you from ever losing, no matter who goes first and what moves your opponent makes. Draughts and chess (or Chinese chess) are games that are more satisfying because they lack this completely predictable completeness. The simplest game which could continue for ever is claimed to be Edward De Bono's L-Game.[8] Each player has an L-shaped token which can be placed anywhere on the small board. After

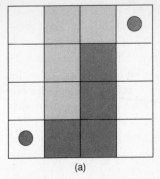 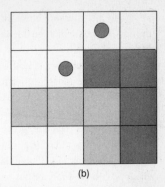

(a)   (b)

Fig. 1.2 (a) The position of the pieces at the start of the L-Game devised by Edward de Bono. A player moves by first repositioning his or her L-shaped piece, and can then move one, or two, or neither of the spots to unoccupied squares. The aim is to prevent your opponent moving their L-shaped piece. (b) A winning position for the light-coloured L, with black to move. The black L cannot now be moved.

placing the L-piece, either one, two, or neither of the black spots may be placed on the empty squares. The aim of the game is to prevent your opponent from moving his L-shape on the next move. The starting positions and a typical winning configuration are shown in Fig. 1.2.

Some games with deceptively simple rules, like John Horton Conway's Game of Life,[9] possess so many developments of great complexity that it is impossible to determine all the possible configurations that could arise. In fact, this game has been shown to share the same level of complexity as the whole of arithmetic. We might wonder whether our investigations of the natural world will eventually be completed in any sense. Perhaps all the laws of Nature might be found, even if all their outworkings might not be listable? Like the perennial noughts-and-crosses addict, would we then cease to be surprised by anything we found in the natural world? In later chapters we shall return many times to look at this question from a variety of different angles.

Those for whom all things are possible

With men this is impossible; but with God all things are possible.
ST MATTHEW[10]

The notion of the impossible has a history bound up with our religious desires. Most human cultures have displayed a desire to worship or acknowledge beings or spirits greater than themselves. These 'gods' are usually credited with superhuman powers: that is what distinguishes them from mortal men and women. Their powers may be exaggerated human ones, or powers that humans do not possess in any measure at all. In the most extreme case the gods may possess limitless powers which enable them to do anything at all and to know everything.

This deceptively simple idea is not without its problems. We can see that it is attractive for the adherents of a particular deity to believe in their god's limitless powers, if only to avoid subservience to the god next door. But looking a little deeper, we see that if their god's actions were limited in some way, then whatever, or whoever, was doing the limiting would have a greater claim to be in control of events than the god. If your god has no jurisdiction over the wind, then the wind has a justifiable claim to be a superior deity. Eventually, someone will appeal to the superior power of the wind.

Although a deity of limited powers has a credibility problem, one of limitless power seems to have far deeper problems of principle. How can there exist a Being for whom nothing is impossible? For whom $2 + 2 = 5$; whose existence can be terminated; who is not bound by the laws of logic? Surely some things must be impossible or chaos and contradiction beckons? If a deity has defining characteristics then there must exist opposites of those attributes which define impossible actions for him or her. Few traditional religions now grapple with these hard questions,[11] yet they are questions that clearly trouble many scientists. The late Heinz Pagels tells how this question was decisive in destroying his early belief in God:

> When I was in high school I remember reflecting on what kind of being God could possibly be—I was curious . . . I also remember asking that if God was all-powerful, could he do things like change the laws of logic? If he could change the laws of logic, then he was a kind of lawless Being incomprehensible to the human mind. On the other hand if he couldn't change the laws of logic, he wasn't all-powerful. These alternatives left me dissatisfied . . . this 'teenage theology' left me with the feeling that either God was not subject to the laws of logic, in which case there was no point thinking rationally about God, or he was subject to the laws of logic, in which case he was not a very impressive God.[12]

Some are content with the notion of a 'miracle', an event which defies the rules by which Nature operates (or, at least, of our experience of them), but none elevate violations of the laws of logic or mathematics to the same evidential status.

Ancient authorities tried to distinguish more finely between actions which were in character and those which were out of character, regarding the latter as logically impossible for a being with the attributes of deity. But these distinctions seem rather slippery to modern ears. Some apologists for the miraculous stress the incompleteness of our knowledge of what is possible in the Universe, and have sought to accommodate God's action in exceptions to the laws of Nature, while others have tried to explain it by our inability to determine the future course of chaotically sensitive situations.[13]

If we look at a religious tradition like the Judaeo-Christian one, we find that God's ability to do the humanly 'impossible' is a defining characteristic. 'To believe only possibilities is not faith, but mere philosophy', as Thomas Browne argued back in the seventeenth century.[14] This feature also serves to establish one of the defining differences between God and mankind: human limits are what fix the great gulf between God and humanity. Thus, when magicians and shamans arise they seek confirmation of their status by demonstrating apparently miraculous powers and by their ability to perform acts which are impossible for the rest of us. They endorse a view of the Universe in which there is a hierarchy of beings whose status rises as the limitations on their actions grow fewer and weaker.

Our religious traditions reveal that restrictions on human thoughts and actions are often *imposed* by the gods. These are not limits which our mortal nature prevents us surpassing: they are like the motorway speed limit rather than the law of gravity. They are presented as taboos that we ignore at our peril. A huge range of human cultures have taboos, whether it be on naming gods, visiting certain places, or counting their populations.[15] Just as earthly rulers distinguish themselves from their subjects by the imposition of constraints upon their behaviour which are not of any obvious benefit to the rulers, except to impress their subjects, so it is imagined that the deity must follow similar practices. The habit of obedience is thought to be a valuable lesson for everyone to learn—a notion that any army sergeant-major will heartily endorse. Thus we see that the notion of impossibility has lodged itself effortlessly at the heart of our religious thinking in many different ways.

The forbidden fruit of the 'Tree of Knowledge of Good and Evil' in the book of Genesis[16] is an interesting example because it entwines two notions that are often separated: forbidden actions and forbidden knowledge. Eating from the Tree of Knowledge was forbidden in order to prevent awareness of some new form of knowledge. The term 'forbidden fruit' has since become a byword for any sort of taboo on human actions.

It is quite common to encounter forbidden actions: our legal systems abound with them. Forbidden knowledge is a more controversial idea. All modern states have secrets and we keep some information concealed from certain people for various reasons—security, confidentiality, financial advantage, malice, surprise, and so on—but there are many who believe that there should be complete freedom of *information* whatever form it takes—as a fundamental human right, like the right to justice and education. This issue has run into controversy with the imposition of restrictions on the Internet and on the attitudes of some governments to the availability of simple encryption programs like PGP ('Pretty

Good Privacy'[17]) which are beyond the means of any government's computer system to break. Alternatively, one can adopt the (British) compromise position that knowledge is not special. Like any human activity or possession (guns, cars, etc.) it may need to be subject to some democratically imposed restrictions for the common good (just as you wouldn't like your credit card PIN number published each day in the papers).

Religious taboos are usually framed in order to maintain the exclusivity of the gods. Some things must be impossible for everyone else if omnipotence is to have any advantage for its possessor. In some Islamic cultures there was a reluctance to produce perfect patterned mosaics because this would trespass into the realm of perfection that is the sole preserve of Allah. Thus, whereas in some religions there are things which humans cannot know because of their finiteness and mortality, in others there are things which they know how to do but must not do, for fear of offending the exclusivity of the gods.

Alan Cromer has argued that the great monotheistic faiths like Islam and Judaism created environments in which science found it hard to develop primarily because they were focused upon deities for whom there was no sense of impossibility:

> Belief in impossibility is the starting point for logic, deductive mathematics, and natural science. It can originate only in a mind that has freed itself from belief in its own omnipotence.[18]

By contrast, the presence of an omnipotent, interventionist being who is unrestricted by laws of Nature undermines faith in the consistency of Nature. A concept of impossibility seems to be a necessary prerequisite for a scientific understanding of the world. This is an interesting argument because it has also been claimed that monotheism provided an environment in which science could flourish because it gave credence to the idea of universal laws of Nature.[19] The decrees of an omniscient deity gave rise to belief in laws imposed on things from outside which govern the workings of the world, in opposition to the idea that the things in the world behaved as they did because of their immanent properties. The distinction is significant. If every stone behaves in a manner dictated by its inward nature, or so as to produce harmony with other stones, then every stone should behave differently and there is little motivation to search for habitual behaviours shared by all moving stones. A feature of this position is that while it is consistent with the growth of abstract science and the concept of externally imposed laws of Nature, it does not ensure it. Although there is strong evidence from ancient China that the absence of a monotheistic view hindered the development of the mathematical sciences and led to a waning of faith in the underlying unity and rationality of Nature,[20] it is not possible to demonstrate that Western science was an inevitable consequence of the Judaeo-Christian and Islamic cultures in the sense that it would not have developed in the absence of their monotheistic beliefs. It may well have been an unexpected by-product of a theistic world-view, but the aims and approaches to the world of these two cultures can be very different. Perhaps, as Oscar Wilde once remarked in a rare moment of seriousness, 'Religions die when they are proved true. Science is the record of dead religions.'[21]

We began this section by introducing the familiar idea of a god who is omniscient: someone who knows everything. This possibility does not immediately ring alarm bells in our brains; it is plausible that such a being could exist. Yet, when it is probed more closely one can show that omniscience of this sort creates a logical paradox and must, by the standards of human reason, therefore be judged impossible or be qualified in some way. To see this consider this test statement:

THIS STATEMENT IS NOT KNOWN TO BE TRUE BY ANYONE.

Now consider the plight of our hypothetical Omniscient Being ('Big O'). Suppose first that this statement is true and Big O does not know it. Then Big O would not be omniscient. So, instead, suppose our statement is false. This means that someone must know the statement to be true; hence it must be true. So regardless of whether we assume at the outset that this statement is true or false, we are forced to conclude that it must be true! And therefore, since the statement is true, nobody (including Big O) can know that it is true. This shows that there must always be true statements that no being can know to be true. Hence there cannot be an Omniscient Being who knows all truths. Nor, by the same argument, could we or our future successors, ever attain such a state of omniscience. All that can be known is all that can be known, not all that is true.

As an aside, we note that the American political scientist, Stephen Brams, has carried out a fascinating analysis of many traditional theological questions relating to God's action in the world, for example the problem of suffering.[22] Brams uses the methods of 'game theory', a branch of mathematics designed to ascertain whether there are optimal strategies for individuals who have different courses of action open to them. The word 'game' is used to describe any situation where two or more participants have a choice of strategies with associated costs and benefits. Brams sought to discover whether we could glean any evidence that the moral nature of the Universe reflects the optimal strategy of an omniscient being. The results were illuminating. Evil and suffering can be inevitable aspects of an optimal strategy to do good. It can turn out the deduction of an omniscient being's existence is logically undecidable if certain strategies are being adopted.

The limitations that this lack of omniscience ensures should not be seen solely in a negative light. Errors and inconsistencies play an important role in our learning process. We learn by our mistakes. If we encounter inconsistencies we re-evaluate the situation as a whole and re-examine the assumptions we have made. It is far from clear to what extent machine intelligence will emulate us in this respect. At some stage in the evolutionary process we began to develop the faculty of imagination. This enabled us to learn about the impossible as well as the possible. Our ability to understand the world thereby increased significantly in scope and speed. Remarkably, we are able to conceive of things that are impossible. Indeed, most of us live our daily lives confident that all manner of impossible things are not merely possible, but actual. Most of us have more interest in the possible than the impossible (this attitude is sometimes called 'pragmatism'); but some people take a greater interest in the impossible. Nor are the latter simply idealists or fantasists. Whole genres of fantastic literature and art have sprung from the challenges posed by linguistic and visual impossibilities.

Paradox

A paradox is truth standing on its head to attract attention.
NICHOLAS FALLETTA[23]

The word 'paradox' is a synthesis two Greek words, *para*, beyond, and *doxos*, belief. It has come to have a variety of meanings: something which appears contradictory but which is, in fact, true; something which appears true but which is, in fact, contradictory; or a harmless chain of deductions from a self-evident starting point which leads to a contradiction. Philosophers love paradox.[24] Indeed, Bertrand Russell once remarked that the mark of good philosophy is to begin with a statement that is regarded as too obvious to be of interest and from it deduce a conclusion that no one will believe.

While some paradoxes may be trivial, others reflect profound problems about our ways of thinking and challenge us to re-evaluate them or so seek out unsuspected inconsistencies in the beliefs that we held to be self-evidently true. Anatol Rapoport, an international authority on strategic analysis—an arena where paradoxical results often result from innocuous beginnings—draws attention to the stimulating role that the recognition of paradox has played in many areas of human thinking:

> Paradoxes have played a dramatic role in intellectual history, often foreshadowing
> revolutionary developments in science, mathematics, and logic. Whenever, in any
> discipline, we discover a problem that cannot be solved within the conceptual
> framework that supposedly should apply, we experience shock. The shock may
> compel us to discard the old framework and adopt a new one. It is to this process
> of intellectual molting that we owe the birth of many of the major ideas in
> mathematics and science. Zeno's paradox of Achilles and the tortoise gave birth to

the idea of convergent infinite series. Antinomies (internal contradictions in mathematical logic) eventually blossomed into Gödel's theorem. The paradoxical result of the Michelson–Morley experiment on the speed of light set the stage for the theory of relativity. The discovery of wave–particle duality of light forced a reexamination of deterministic causality, the very foundation of scientific philosophy, and led to quantum mechanics. The paradox of Maxwell's demon, which Leo Szilard first found a way to resolve in 1929, gave impetus more recently to the profound insight that the seemingly disparate concepts of information and entropy are intimately linked to each other.[25]

Visual paradox

> *You arrive at the truth by telling a pack of lies if you are writing fiction, as opposed to trying to arrive at a pack of lies by telling the truth if you are a journalist.*
>
> MELVIN BURGESS[26]

The divergence of the artistic and scientific pictures of the world has been made most striking by the focus of twentieth-century artists upon abstract images and distortions of the everyday picture of the world. One of the most extraordinary consequences of human consciousness is the ability it gives us to imagine things which are physically impossible. By this device we can explore reality in a unique way, placing it in a context defined by impossible events. In this way we are able to create resonances of meaning and juxtapositions of ideas which are mind-stretching and stimulating. This we find appealing and novel. Some individuals devote their lives to this activity, creating and appreciating these alternative realities in a host of different media. The affinity that our minds possess for this activity is almost alarming. The sudden appearance of sophisticated computer simulations of alternative realities and the ready availability of computer games which are indistinguishable from direct human activities have revealed how seductive such experiences are to young people. They offer a huge range of vicarious experience without the need to leave the comfort of one's chair. Perhaps the appeal of these virtual adventures is telling us something about the untapped potential within the human mind which is so little used in the cosseted activities of everyday twentieth-century life. We have begun to use the computer interactively in education, but with little imagination so far. I suspect there is a great opportunity here to teach many subjects—especially science and mathematics—in an adventurous new way. Even a mundane computer-based activity, like word processing, has done more than make writing and editing more efficient: it has altered the way in which writers think. Writers used to write because they had something to say; now they write in order to discover if they have something to say.

The representation of the impossible has become a prominent part of the modern artistic world. This takes several forms. The graphic style of Maurits Escher[27] employs a form of precise drawing which seeks to deceive the viewer into believing that he has entered a possible world which, on closer scrutiny, turns out to be inconsistent with the nature of space in which we live. Escher likes impossible objects which we could define as two-dimensional images of apparent three-dimensional objects which cannot exist as we have interpreted them: that is, they cannot be constructed in three-dimensional space.

The three-dimensional interpretation of these images is a different matter. The eye is led to build up different local pictures which, ultimately, cannot be combined into a single consistent visual scenario. In modern times impossible objects were drawn first by Oscar Reutersvärd.[28] In 1934 he drew the first known example of an impossible tribar (Fig. 1.3a). Escher created the first impossible cube in 1958. The tribar was rediscovered in 1961 by Lionel and Roger Penrose, who introduced the never-ending staircase (Fig. 1.3b)[29] Escher employed these in his famous drawings *Waterfall* (1961) and *Ascending and descending* (1961).

There are a number of curious older examples of this genre which have been recognized retrospectively. Hogarth's engraving on copper *False perspective* (1754)[30] is a beautiful example (Fig. 1.4). It was drawn by Hogarth to exaggerate the mistakes of inept draughtsmen. He labels the picture, 'whoever makes a Design without the Knowledge of Perspective will be liable to such Absurdities as are shewn in this Frontispiece'.

In 1916, Marcel Duchamp created an advertisement for the paint manu-facturers Sapolin.[31] The bed frame incorporates a tri- and four-bar structure (Fig. 1.5). The original, entitled *Apolinère enameled*, is now in the Philadelphia Museum of Art.

The famous Italian architect and engraver Giovanni Piranesi (1720–78) produced a sinister collection of designs for a series of labyrinthine dungeons between 1745 and 1760. These fantastic creations depicted impossible networks of rooms and stairways. His working diagrams reveal that he deliberately set out to create impossible configurations.[32]

Breughel's *The Magpie on the Gallows* (1568) deliberately makes use of an impossible four-bar. Unintentional impossible objects can be found at very early times. The oldest known example dates from the eleventh century.[33]

These impossible figures reveal something more profound than the draughts-man's skill. They tell us something about the nature of space and the workings of the brain's programming for spatial analysis. Our brains have evolved to deal with the geometry of the real world. They have defence mechanisms to guard against being deceived by false or ambiguous perspective. In such a dilemma the brain changes the perspective adopted every few seconds as an insurance against having made the wrong choice. A common example is the Necker cube (Fig. 1.6), which seems to flit back and forth between two different orientations.[34]

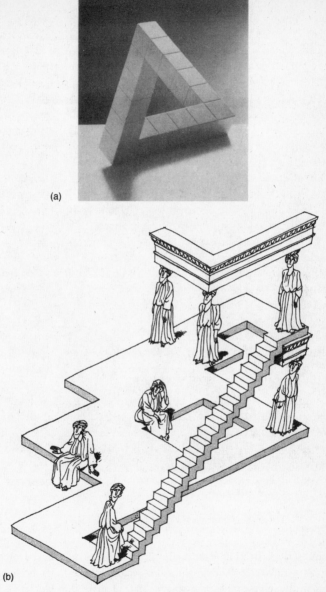

(a)

(b)

Fig. 1.3 (a) the first modern drawing of an impossible object, a tribar composed of nine cubes, was made by the Swedish artist Oscar Reutersvärd in 1934 (© DACS 1998). (b) A continuous staircase shown in the drawing *Caryatids*, also by Reutersvärd, with human figures added by Bruno Ernst to emphasize the spatial dissonance (© DACS 1998).

Fig. 1.4 William Hogarth's copper engraving, *False Perspective* (1754).

Surrealist works of art have other aims. They stimulate the mind by forcing it to evaluate and accommodate situations which it believes to be logically impossible. By representing an impossible state of affairs they lay claims upon our attention in memorable ways. By this means, they establish themselves as something quite distinct from the real world of experience, and not merely an accurate copy of it. A classic example is provided by a picture like Magritte's *Le*

Fig. 1.5 Marcel Duchamp's advertisement *Apolinère enameled* (1916/17). (Philidelphia Museum of Art: The Louise and Walter Arensberg Collection. © ADAGP, Paris and DACS, London 1998.)

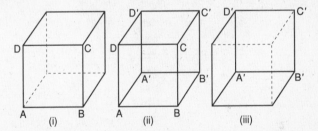

Fig. 1.6 The Necker Cube, with all lines solid, is shown in the centre (ii). On either side, (i) and (iii), we show alternative visual interpretations of it in which the cube appears differently oriented. The eye makes rapid shifts between the two interpretations (i) and (iii). Solid lines are in the foreground; dotted lines in the background.

Château des Pyrénées of a gravity-defying castle in the air (Fig. 1.7).[35] Perhaps we like imaginary worlds that are impossible because their very impossibility reinforces the appeal of artistic representations of strange environments and circumstances which we can experience safely. They allow us to enter environments which are dangerous, in the sense that they could not possibly be part of our (or anyone's) experience, but without real risk. They are an extension of the phobophilia that attracts us to the ghost train or the horror film.

Fig. 1.7 René Magritte, *Le Château des Pyrénées* (1959). (Reproduced courtesy of the Israel Museum, Jerusalem.) © ADAGP, Paris and DACS, London 1998.

Much has been made of the way in which geometrically distorted pictures began to appear at a time when physicists first began to appreciate the physical relevance of geometries other than Euclid's. Pioneering cubists like Picasso always denied that scientific developments motivated them in any direct way.[36] Escher, on the other hand, seemed to appreciate the studies that mathematicians made of other geometries. Indeed, his work may even have stimulated some explorations of new tessellations of space.[37]

There is also a complementary literary style which trades on impossibility and paradox. The greatest early exponent of this was probably the Victorian surrealist Lewis Carroll. We see its more eclectic and fantastic manifestations in the short stories of Jorge Luis Borges, and others.[38] The conjuring up of worlds that don't quite fit remains a strangely attractive creative activity: the only way to be truly original.

The interesting feature of all these examples is the way in which they show our recognition of the impossible. The impossible is not necessarily something that lies outside our mental experience even if it falls outside our physical experience. We can create mental worlds which are quite different from the one we experience. Indeed, some people clearly relish these images of impossible worlds as much as any that could be made of this one.

Linguistic paradox

The supreme triumph of reason is to cast doubt upon its own validity.
MIGUEL DE UNAMUNO

Impossible figures are examples of visual paradoxes, or perhaps we should say inverted paradoxes. A paradox is usually something which, although seeming to be false, is in fact true. Impossible figures are things which, despite seeming true, are in reality false. We might have expected that our reaction to paradox would be one of confusion or aversion. Paradoxically, it is apparently quite the opposite. We enjoy paradox: it lies at the heart of many forms of humour, stories, pictures, and a host of well-appreciated quirks of human character.

Paradoxes spun for amusement have a habit of subsequently proving deeply profound. History is strewn with examples. Zeno's paradoxes have stimulated our understanding of the infinite.[39] Zeno was Greek philosopher of the fifth century BC who is best known for these paradoxes, which appear to show that motion is impossible. His most famous example is that of the race between Achilles and the tortoise. Suppose that the tortoise is given a 100-metre start but Achilles runs a hundred times faster than the tortoise. While Achilles runs 100 metres, the tortoise covers 1 metre; while Achilles runs 1 metre, the tortoise covers 1 centimetre; and so on, for an infinite number of steps. As a result Achilles will never catch the tortoise! The problem can be resolved if we recognize that although an infinite number of instants of time will have elapsed before Achilles catches the tortoise, it is not necessarily true that an infinite number of instants of time must add up to make an infinitely long time.[40]

In modern science the term 'paradox' is usually reserved for a counter-intuitive finding that is believed to shed light upon something fundamental. Thus we have the 'twin paradox' of relativity,[41] Schrödinger's 'cat paradox',[42] the 'Einstein–Podolsky–Rosen (EPR) paradox',[43] the 'Klein paradox; of quantum field theory,[44] and the paradox of 'Wigner's Friend' in quantum measurement.[45] These 'paradoxes' may be created by some incompleteness of our knowledge of what is going on, either at the level of the theory supposed to describe it, or in the specification of the state of affairs that is observed. Alternatively, they may appear paradoxical only because our expectations are simply wrong and derive from very limited experience of reality (as in the case of the 'twin paradox'). We can expect that further development of our understanding will either resolve the apparent paradox or reveal that there is in fact no paradox.

Linguistic and logical paradoxes are not like this at all. They are simple enough for everyone to appreciate. They affect the very tools that we use to think about everything and are therefore more deeply disturbing. Logic seems to be the final stop for human thinking. We can reduce science to mathematics and mathematics to logic, but there seems to be nothing to which we might reduce logic. The buck stops there.

Logical paradoxes have a long history. The most famous is repeated by St Paul in his Epistle to Titus when he remarks that 'all Cretans are liars, one of their own poets has said so.'[46] This is the Epimenides (or 'Liar') Paradox.[47] For centuries such paradoxes appeared to be little more than isolated curiosities that could safely be ignored because they never seemed to arise in situations of practical importance. But during the twentieth century their importance has grown into something fundamental. They are consequences of logical structures which are complex enough to permit self-reference but arise when we are insufficiently careful to distinguish statements made in a particular language from those made in another language. Far from confining the linguistic paradoxes to the world of triviality, this distinction ends up by giving them a central role in formal proofs of the logical incompleteness of logical systems.

One of the most notable modern thinkers to be troubled by paradoxes was the philosopher Bertrand Russell, who wrote about his discovery, in June 1901, that logic contains a fundamental inconsistency. Subsequently, it became known as the 'Russell Paradox'.

> It seems to me that a class sometimes is, and sometimes is not, a member of itself. The class of teaspoons, for example, is not another teaspoon, but the class of things that are not teaspoons, is one of the things that are not teaspoons ... [this] led me

to consider the classes that are not members of themselves; and these, it seemed, must form a class. I asked myself whether this class is a member of itself or not. If it is a member of itself, it must possess the defining properties of the class, which is to be not a member of itself. If it is not a member of itself, it must not possess the defining property of the class, and therefore must be a member of itself. Thus each alternative leads to its opposite and there is a contradiction.

The most memorable formulation that Russell gave to this difficulty of the set of all sets that are not members of themselves was to tell us of a town in which there is a barber who shaves all those who do not shave themselves. Who shaves the barber?[48] What worried Russell so much about this paradox was its infiltration of logic itself. If any logical contradiction exists it can be employed to deduce that anything is true. The entire edifice of human reasoning would fall. Russell was deeply pessimistic of the outcome:

> Every morning I would sit down before a blank sheet of paper. Throughout the day, with a brief interval for lunch, I would stare at the blank sheet. Often when evening came it was still empty . . . it seemed quite likely that the whole of the rest of my life might be consumed in looking at that blank sheet of paper. What made it more annoying was that the contradictions were trivial, and that my time was spent in considering matters that seemed unworthy of serious attention.

Later, we shall discover that these seemingly innocuous linguistic paradoxes revealed the presence of profound problems for the whole of logic and mathematics, showing there to be a trade-off between our ability to determine whether statements are true or false and our ability to show that the system of reasoning we are employing is self-consistent. We can have one or the other, but not both. We shall find that there are limits to what mathematics can do for us: limits that are not merely consequences of human fallibility.

Limits to certainty

> *There is a theory which states that if anyone discovers exactly what the Universe is for and why it is here, it will instantly disappear and be replaced by something even more bizarre and inexplicable. There is another theory which states that this has already happened.*
>
> DOUGLAS ADAMS[49]

The linguistic and logical paradoxes we have been considering go back thousands of years to the ancient Greeks. But in modern times we have encountered a different breed of paradox: one that governs what we can do rather than simply what we can say. During the first quarter of the twentieth century the twin discoveries of relativity and quantum theory revealed that there are unexpected limits to what can happen under extreme conditions. As experiments and theoretical investigations probed towards the frontiers of small sizes, large sizes, high speeds, very strong gravity fields, very high energies, and very low temperatures, they invariably encountered an unexpected limit on what could be done or what could be known about the state of the Universe. These were unexpected because they ran counter to what was predicted by simply extrapolating our experience of the laws of Nature from moderate laboratory conditions to unfamiliar environments. Two of these, the limits to measurement that the quantum nature of matter requires, and the cosmic speed limit imposed by relativity, are now foundation stones of our understanding of the physical world.

One of the most enthusiastically popularized areas of science in recent years has been quantum theory.[50] This is somewhat surprising to insiders because nothing new has happened in the subject. The theory was completed long ago. All the subsequent journalistic interest has been in its interpretation. Part of the mystique of quantum theory is that it combines amazing experimental success with a panoply of contrary-to-common-sense assertions about the world. Its domain is the small scale of atoms and their aggregates. Its potential to surprise arises because our familiar intuition about the behaviour of moving objects is gleaned from our experience of relatively large objects.

Quantum theory teaches us that all objects possess a wave-like aspect. This aspect is wave-like in the sense of a crime wave rather than a water wave. That is, it is a wave of information. If a neutron wave passes through your detector it tells you that a neutron is more likely to be detected there. The wavelengths of these matter waves are inversely proportional to their physical sizes. When an object has a quantum wavelength that is larger than its physical size it behaves in an overtly quantum fashion; when its wavelength is smaller than its size it behaves in the classical Newtonian fashion. Thus, typically, very large objects like you and me are said to behave 'classically', whereas small objects like elementary particles behave 'non-classically' or quantum-mechanically. Classical behaviour is just the extreme limit of quantum behaviour when the physical size of an object gets much bigger than its quantum wavelength.

One of the curiosities of the quantum realm is that some classically impossible things become possible and some classically possible things turn out to impossible. For example, in classical Newtonian science we assumed that it was possible to know simultaneously both the position and motion of a particle with complete accuracy. In practice, there might be technological limits to the accuracy with which this could be done, but there was no reason to expect that there was any limit in principle. On the contrary, we would expect that ever-improving technology would enable this accuracy to keep getting better, just as it always had done. But quantum mechanics teaches us that even with perfect instruments it is impossible to measure the location and velocity of a body simultaneously with an accuracy better than some critical limit defined by a new constant of Nature, called Planck's constant. This constant, and the limiting accuracy it prescribes, is one of the defining characteristics of our Universe. It will place just the same limits on what physicists in the Andromeda Galaxy can do as it does for physicists on Earth.

The limit on our accuracy of measurement is known as Heisenberg's Uncertainty Principle. One heuristic way of understanding why there should be such a limit is to recognize that measurement requires an interaction of some sort with the state that is being measured: the smaller the thing being measured, the greater the impact of the measurement process. Eventually, that impact supersedes all information about the unperturbed state. The quantum picture of reality thus introduces a new form of impossibility into our picture of the world. This impossibility replaces a past belief in unrestricted experimental investigation of Nature which was based upon a misconception of what existed to be measured. There is a more accurate way to view the Heisenberg Uncertainty. It is not, as in our simple heuristic example, that there is a definite reality that we are unable to capture because measurement requires inter-vention. This suggests that we might be able to calculate what the effect of a particular intervention would be and allow for it in advance. Rather, the Uncertainty Principle is telling us that in the quantum realm, where dimensions are sufficiently small, certain complementary pairs of concepts, like position and velocity, or energy and time, can coexist only with a limited sharpness that Planck's constant dictates. The concepts referred to are classical concepts and there is a limit to their application. It is only because we had assumed (wrongly) that there was no limit in principle to our ability to measure all measurable quantities that we are shocked by Heisenberg's Uncertainty Principle and think of it as some sort of a limit on what we can do. Heisenberg teaches us that the scientist is not like a birdwatcher in a perfect hide. Observing the world necessarily couples us to it and influences its state in ways that are only partially predictable or knowable.

Heisenberg's Principle has had a widespread impact upon human thinking about certainty and knowledge.[51] It is a prominent feature of many discussions of the interface between science and religion because it provides a ready-made guarantee that there must always be a gap for a God-of-the-gaps argument to fill. In general, the tenor of this discussion welcomes rather than despairs of the ignorance that Heisenberg guarantees. There have occasionally been attempts to find mental consequences of Heisenberg uncertainty, but the general opinion is that the effects are too small on the scale of neurones to have any significant effect upon the human thinking process.[52] Natural selection would certainly lead us to expect this: if significant irrationality was created by the limits set by the Uncertainty Principle, then there would have been a significant reduction in the chance of survival. Neuronal networking that evolved on a scale large enough to avoid significant quantum uncertainty would have been more adaptive than varieties on smaller scales susceptible to quantum uncertainties.

The fact that our world possesses quantum uncertainty at all is a consequence of the fact that Planck's constant is not equal to zero. We do not know why it takes the exact non-zero value that it does. If it were larger than it is, then larger objects would display strong wave-like attributes. The famous 'Mr Tompkins' stories by the late George Gamow attempted to explain some aspects of quantum reality by showing what the world might be like if Planck's constant were so large that everyday objects became overtly wave-like in character.[53]

The classical Newtonian laws that govern how bodies move prescribe rules of cause and effect. If a body is subjected to a certain force it will move with a definite acceleration. These laws enable the path taken by a body acted upon by forces to be calculated exactly if we know its starting state. In this way we can calculate the orbit of a planet around the Sun. Thus we see that laws of Nature involve the idea that certain motions are impossible; that is, if they occurred they would violate the laws of motion or some attendant principle like the conservation of energy. In quantum mechanics this picture changes in an extraordinary way. Quantum mechanics gives no exact predictions for the future location and speed of motion of an object given its starting state. It gives only probabilities that it will be observed to be at some location with some velocity. If the moving object is large (in the sense described above) then those probabilities will have a negligible spread and for all practical purposes (a probability almost exactly equal to 100 per cent certainty) the position and velocity of the object will be as predicted by Newton's laws. If, however, the object is small enough for its wave-like character to be significant, there may be an appreciable probability for it to be found in a state of motion that is impossible according to Newton's laws. Such states are frequently observed. They serve to distinguish the behaviour of the microscopic world from that of everyday experience. In quantum mechanics anything might be observed with some probability—although that probability might be vanishingly small.

A cosmic speed limit

The simplicities of natural laws arise through the complexities of the languages we use for their expression.

EUGENE WIGNER[54]

In the early years of the twentieth century Albert Einstein completed a picture of Nature to which many other scientists had contributed without seeing so deeply and clearly what all the pieces added up to produce. Einstein showed that Newton's laws of motion broke down when applied to the motion of bodies moving at high speed. They were just a good low-speed approximation to a more general set of laws which governed motion at all possible speeds. But what do we mean by 'high' and 'low' speeds? Remarkably, Nature tells us in a way that involves no subjective judgements and no reference to our own motion. All speeds are to be judged relative to the speed of light in empty space. This speed, equal to 229,792,458 metres per second (about 186,000 miles per second) is a cosmic speed limit.[55] No information can be transmitted by any means at a speed exceeding this value. (Note that light travels more slowly through a medium than through empty space and it *is* possible to transmit information through a medium at a speed faster than the speed of light in that medium so long as it travels more slowly that the speed of light in empty space.[56] Newton's laws of motion predict no such speed limit (information is transmitted instantaneously) and they lead to incorrect predictions about the world when applied to the motion of particles moving at speeds close to that of light. This is the regime of 'high-speed', or relativistic, motion.

The fact that there is a limit to the speed at which information can be transmitted in Nature has all sorts of unusual consequences. It is responsible for our astronomical isolation. The enormous times needed to send or receive light or radio waves from other star systems in the Universe is a consequence of the finite speed of light. It is also responsible for our own existence in ways that may not be at first obvious. If the speed of light were not finite, then radiation of all sorts would be received instantaneously after it was emitted, no matter how far away its source. The result would be a reverberating cacophony. We would be dramatically influenced by signals from everywhere. Instead of local influences dominating over far distant ones, we would be affected instantaneously by changes occurring on the other side of the Universe. The impossibility of transferring information faster than the speed of light makes it possible to discriminate and organize any form of information.

Our world is governed by relativity because the speed of light is finite. We do not know why the speed of light takes the specific value that it does in our Universe. If it were much smaller, then more slowly moving objects would suffer the distortions of space and time that arise as the speed of light was approached; less energy would be available when matter was annihilated in nuclear reactions; light would interact more strongly with matter; and matter would be less stable.

Again, we see a twofold evolution of our ideas about impossibility and its consequences. Before Einstein, the Newtonian picture of the world placed no limit on the speed at which light or any other form of information might be transmitted in the Universe. But the connection between that assumption and other aspects of the structure of the Universe was not recognized. In reality, a Newtonian universe was impossible. It was too simple to accommodate light.

After Einstein, we are faced with the recognition that faster-than-light information transmission or space travel is in general impossible but this impossibility is what makes the self-consistency of the laws of Nature possible.

Summary

I dreamt I died and went to heaven, and Saint Peter led me into the presence of God.
And God said 'You won't remember me, but I took your Quantum Mechanics
Course in Berkeley in 1947.'

ROBERT SERBER[57]

We have begun to explore some of the ways in which the notion of impossibility lies at the root of many flowerings of the human imagination. We have taken some snapshots of different parts of our cultural development which have made important use of the concept of impossibility, both as a constraint on human actions and by way of contrast with the concept of a Being for whom nothing is an impossibility. Impossibility has played a stimulating role in art through the creation of impossible figures. In philosophy, paradoxes have been of persistent interest, leading to profound new considerations of the problems of the infinite and the nature of language, truth, and logic. Finally, we saw two examples of developments of our understanding of the physical Universe which showed us that there were unsuspected limits on what we can measure and how fast we can transmit information. The development of complex descriptions of the workings of the physical world seems to lead inevitably to theories that know their own limitations: that predict that they cannot predict.

These excursions lead us to begin to look more closely at the types of limit that we might encounter in our quest to understand the Universe, to consider whether we can expect to keep on progressing, and what 'progress' means.

CHAPTER 2

The hope of progress

> You've got to ac-cent-tchu-ate the positive
> Elim-my-nate the negative
> Latch on to the affirmative
> Don't mess with Mister In-between.
>
> JOHNNY MERCER

Over the rainbow

> The irony of life is that it is lived forward but understood backward.
> SØREN KIERKEGAARD

We can look back over a century of unprecedented progress in most areas of practical achievement. Machines, medicines, education, computer systems, transport, . . . the roster of achievements seems endless and relentless. Progress is undeniable, but what of the *rate* of progress. Is it accelerating or decelerating? Will our knowledge of Nature continue to grow? Or could it eventually slow to a trickle?

During the past thirty years, science has steadily mopped up lots of problems that were opened up by new technologies. New knowledge has invariably meant new gadgets and ways of transferring information which require ever-decreasing amounts of time and energy. But will new knowledge always have new practical consequences? Or will the frontiers of the doable lag further and further behind those of the conceivable?

Present theories of physics lead us to believe that there are surprisingly few fundamental laws of Nature. Nevertheless, there seems to be an endless array of different states and structures that those laws permit—just as there are a very small number of rules and pieces defining a game like chess, yet an endless number of different games that could be played out.[1] Any unfound forces must be extremely weak or severely constrained in their effects, perhaps confined to very short distances or to influencing the behaviour of very rare ephemeral entities. Physicists are fairly confident that they are not missing something in between the forces that they have already found.[2] When it comes to the outcomes of those laws there is no comparable degree of confidence. There is a steady flow of new discoveries and a growing appreciation of how complex organized structures come about and evolve in tandem with their environments. This trend could be just that—a trend—which runs its course, culminating in a full understanding of all the varieties of complexity that can exist. We might just be living in the Golden Age of complexity studies[3] in the way that the 1970s and 1980s were a Golden Age for elementary particle physics. Experimental science is based upon discoveries, and you can only discover America once—as the Vikings would have told Señor Columbus.

Some scientists and philosophers have taken the view that science as a whole has experienced a Golden Age that will eventually draw to a close. Truly new discoveries will become harder and harder to make; minor variations will become tempting targets; deeper understanding will require greater and greater efforts of the imagination to achieve; and a wider grasp of the structure of systems of huge complexity will require more and more powerful computers. The seam of gold that is useful science may one day be mined out, leaving only a few nuggets to be uncovered here and there by ever-increasing effort. Of course, we may not realize that the mine is exhausted; no banner will appear in the sky to tell us that further fundamental advances will require a huge leap for Mankind, rather than a gradual shuffle. The demise of science may come not with a bang but a whimper. The financial cost of unearthing new knowledge may ultimately place too great a drain on scarce human resources. No potential benefit will outweigh the costs of investigation.

Even if this pessimistic scenario is not haunting our future, just contemplating it can help us focus on reality more clearly. The cost of scientific investigation has already become a political issue. How much of the GNP of a country should be spent on scientific investigation with little or no prospect of practical advantage or technical spin-off? How indirect can the benefits of science be and still be counted as benefits that derive from it? In this chapter we shall look at some provocative modern opinions about scientific progress before casting a look back at the prognostications of past prophets who wondered if progress was coming to an end at the turn of their own century. Not content with generalities, they often highlighted scientific problems that they thought would never be solved. Their worries were sometimes very similar to our own.

Scientists alone do not dictate the future course of science. When their activities become very expensive and have no direct technological or military relevance to the state, then their continued support will be determined by other great problems that confront society. If there are climatic problems, then meteorologists and space scientists will be looked upon more favourably by government funding agencies than elementary particle physicists or metal-lurgists. In the future, we might expect that the development of what we will call the 'problem sciences'—those studies needed to solve the great environmental, social, and medical problems that threaten humanity's continued existence and well-being—will be thrust increasingly to the limelight and voted abundant resources. Throughout human history, the threat and existence of war injected urgency and focus into special areas of science and mathematics. In the future that state of urgency may focus our attentions upon the by-products of our own past actions and the impact of untoward climatic and ecological trends in the natural world. Over very long periods of time the low-risk disaster becomes a certainty unless it is constantly guarded against.

Increasingly, it appears that 'advanced' societies—those that have extensive investment and reliance on science and technology—tend to create other internal problems, tensions, and expectations that are expensive to meet. Those that have the wherewithal to fund scientific research invariably have many other calls on their resources. Nor do these calls derive solely from the need to repair careless mistakes. Success can be costly as well. We continually find new medical treatments for conditions that were once untreatable. Yet, the costs of implementing them on a large scale could well prove ruinous to society. The costs of maintaining private and public medical care continue to grow as a result of the cost of more sophisticated treatments and the eradication of illness that were once fatal in late middle age. Every systematic medical success over progressive illness provides a new set of survivors whose encounter with the next affliction of older age will create a new social challenge.

One hope for sustained scientific progress may be the development of computer systems with new levels of miniaturization, speed, and complexity. Pure science projects that promote the development of these new technologies will play a starring role in the future. This computational dividend from exploratory fundamental science is something that we are already familiar with from past 'big science' projects. One of the greatest benefits of the early US space programme was not specimens of Moon rocks, but the rapid advance of large and reliable real-time computer systems. More recently, the Internet worldwide computer network is something that emerged from CERN, the European Centre for Particle and Nuclear Physics.

The success of science has elevated its activities to a new level of size and complexity. 'Big science' means international collaborations of hundreds of scientists, budgets running into hundreds of millions of pounds, and investigation times that can exceed the creative lifetimes of the central participants. One by one, the various sciences will reach a stage where they wish to move forward by embarking upon a vast project in order to join the big-science league. The asymptotic attraction of this type of collective project is the hallmark of a certain type of maturity in a physical science, where there is a successful central theory which is able to make use of huge amounts of data and vast facilities for computational analysis. Physicists were the first to focus like this (on particle accelerators), then astronomers (on the Hubble Space

Telescope), and now biologists (on the Human Genome project). Others will surely follow. The science budgets of most countries have already had to come to terms with scientific activities that were once inexpensive ventures requiring little more than a few test tubes, books, chemicals, home-blown glassware, and low-tech equipment, but now need large computer systems, spectrometers, electron scanning microscopes, small accelerators, and other very expensive pieces of hardware, with formidable running costs, that require frequent upgrading to keep their users at the cutting edge of research worldwide.

The relentless desire for progress which has led to these never-ending demands for money and resources is something deep-rooted in our make-up. Perhaps it is not mysterious. We are the products of a long evolutionary history that has selected for those traits that survive best. An ability to change our environment and so fashion our own ecological niche has enabled us to outstrip other species and survive on every part of the Earth's surface. The harder the competition, the greater the pressure to gain a marginal advantage by the adoption of some innovation. Progressives will have been better adapted to survive in changing environments than conservatives. Progressive activity seems manic at times and creates all sorts of problems but, like growing old, it's not so bad when you look at the alternatives.

Today, most inhabitants of the Western democracies live luxuriously when compared with the lot of their distant ancestors. One might worry that as comfort increases so the incentive to innovate diminishes. Looking forward we might wonder whether the direction in which technological societies are moving—creating less work, longer lives, and greater leisure—might eventually remove incentive and desire to innovate in science and technology. Governments try increasingly to create 'economic environments' which stimulate and reward innovation in order to counter the growth of a dependency culture at the bottom of the scale and a lethargic culture at the top. How will things move in the long term? Creativity might find itself channelled into other areas, as individuals prove to have an unsuspected susceptibility for being sucked into virtual electronic realities and other high-tech amusements. Alternatively, apathy might become endemic. The challenge is not dissimilar to that in the face of being told that you are going to live for ever. Do you rush out to begin the first in a never-ending sequence of new careers, or do you lie back in the knowledge that there is world enough and time to do everything *mañana*.[4] The social analyst José Ortega y Gasset saw this same division when he observed that

> the most radical division that is possible to make of humanity is that which splits it into two classes of creatures: those who make great demands of themselves, piling up difficulties and duties and those who demand nothing special of themselves, but for whom to live is to be every moment what they already are, without

imposing on themselves any effort towards perfection, mere buoys that float on the waves.[5]

And we all know someone of each sort.

While this might be an acute evaluation of a twofold division between human personalities, one must be careful about what it is applied to. It is easy to talk of 'human society', or 'scientists', as though each were a single individual. They are nothing of the sort. Rather, they are whole populations of individuals displaying a wide range of different motivations and beliefs. Those motivations might well cluster around two opposite poles, but in any society there would still be a spectrum of different motivations and beliefs that would offer the prospect of quite different futures from any present mixture of the two.

The voyage to Polynesia via Telegraph Avenue

In Italy for thirty years under the Borgias, they had warfare, terror, murder, bloodshed. They produced Michelangelo, Leonardo da Vinci, and the Renaissance. In Switzerland, they had brotherly love, five hundred years of democracy and peace, and what did they produce? The cuckoo clock.

ORSON WELLES[6]

The increasing cost of pushing back scientific frontiers might lead to a growth in the philosophical analysis of science and the discussion of unanswerable 'meaning-of-life' questions like 'how did the Universe begin?'. In this way, the hard core of science might be mined out, leaving only a superficial veneer of questions about which one can have opinions but not testable answers. The view that science might bring about its own lugubrious demise was first aired in 1969 by the distinguished biologist Gunther Stent, then of the University of California at Berkeley, in his book *The coming of the Golden Age*.[7] His argument has recently been rediscovered and reiterated by the American journalist John Horgan in his book *The end of science*.[8]

Stent thought that science was reaching the end of the road—but not because it was getting too expensive. He thought that the great discoveries had been made and science was heading towards a future of baroque elaboration, subjectivism, and introspection already to be found in many of the creative arts. History teaches us that the ancients harked back to a mythical Golden Age when a privileged race of mortal men lived on Earth in a state of paradise. According to Greek legend, this state of earthly bliss ended when Pandora lifted the lid of her box and released a host of previously unknown evils into the world. The Golden Age was then succeeded by a decline in lustre, through Silver, Brass, and Heroic Ages, until we reached the present Iron Age of labour and sorrow in which Mankind reaps the bitter harvest of the gods. Jewish tradition has a similar, more familiar, story of decline and fall from a realm of Edenic bliss to a world bubbling with toil and trouble.

Stent argued that this mythological picture needs turning on its head. Our scientific Golden Age is not in the past; events point to a Golden Age that is upon us *now*. The most significant feature of this contemporary Golden Age is not the lustre of its achievements but the fact that it marked the culmination of the rapid rise of science. Stent's contemporaries exhibited the signs of having pretty much got where they were going to go. So, what put the brakes on?

Stent did not see the end of science drawing nigh primarily because the readily soluble problems had dried up. Paradoxically, he saw the demise of science as a consequence of its own success in sustaining an unprecedented increase in living standards, social well-being, and security following the deprivations and horror of successive world wars. Science, if successful, tends to bring about social conditions in which the psychological motivations needed to manipulate the natural world for advantage are allowed to atrophy. He wrote,

> I shall try to show that internal contradictions—theses and antitheses—in progress, art, science, and other phenomena relevant to the human condition make these processes self-limiting; that these processes are reaching their limits in our time and that they all lead to one final, grand synthesis, the Golden Age.[9]

This state of affairs is compared to the characteristic history of the South Sea Islands. They were settled by an adventurous race of seafarers who set out from South-East Asia across the Pacific Ocean three thousand years ago in tiny open boats in search of a better place to live. Over the next two thousand five-hundred years, motivated by the search for food and land, they spread out and colonized all the habitable Pacific islands. But when that process was complete, some four hundred years ago, things began a downward spiral. In the face of fertile lands and the abundant harvest of the sea, the spirit of adventure decayed, hedonism grew, intellectual endeavour languished, and the creative arts of the past were left to fade and die.[10] In this sad history of Polynesia, Stent saw the apathetic consequences of a decline of the human 'Faustian' spirit which desires to subdue the environment in new ways,

> the 'threat' of leisure was met at least once before by simply and easily abandoning the gospel of work. It shows that people will not necessarily go stark, raving mad when, in a background of economic security, most of them no longer have much useful employment. The Vikings of the Pacific must have started with a strong Faustian bent, but by the time Captain Cook found them, Faustian man had all but disappeared . . .[11]

In judging these analogies, one must remember Stent's situation. He was writing in Berkeley in 1969, soon after the great student demonstrations by the

Students' Free Speech Movement (which sparked similar protests elsewhere in the world). In Berkeley, there was much soul-searching by scholars and university administrators about the causes and long-term significance of these unprecedented student protests. At the very least, a large segment of American youth had collectively altered their opinions about what were worthwhile goals in life. The American Dream had turned into the American Nightmare. Shaken by this change of direction, Stent thought that American youth had given up the search for knowledge and would never return to it. It was the nature of their protests, rather than what they were protesting about, that depressed him most. They were seen as anti-rational and anti-success. In short, they were anti-progress. The long-term future of rational enterprises like science did not look rosy from the Faculty Club on the Berkeley campus. The close links between science and scientists at Berkeley and the American military (the Livermore weapons laboratory, directed by Edward Teller, was just 45 minutes' drive away, and formally part of the University) did not bode well for the future either. Science was decelerating because of radical social change rather than from any exhaustion of its subject matter.

Stent's thinking was much influenced by the nineteenth-century 'philosophers of progress' who thought that they had found an objective measure of human progress by charting the scope of our power to manipulate the natural world.[12] Following their lead, Stent thought that our evolutionary history had endowed us with an instinct for manipulating and controlling our environments. We can pass it on more rapidly by processes like education, especially of young children, than by the painfully slow process of genetic inheritance, and it is an instinct that increasingly influences the development of industrialized societies. Moreover, when we succeed in manipulating Nature in a manner that is optimal for ourselves, we are 'happy'. But, as society has become more affluent in the post-war years, the social conditions needed to inspire this manipulation began to fade away. The beatnik generation were the first to be raised in conditions of relative prosperity. The economic security of Stent's students had eroded the desire to progress in the way that had been second nature to their predecessors, who experienced the Depression or the hardships of immigration from conditions of poverty or persecution.

When we come to explore the growth and possible limits to technological progress we shall look again at Stent's arguments. Stripped of the specifics of mid-1960s Berkeley, his argument is simply that progress is self-limiting. Because the primary inspiration for progress is a psychological desire to shape our environment and control our futures, the more successful we are in this respect, the more affluent and secure our existence will be, and the less will be our need and desire for further progress. From our vantage-point in time, Stent's prognostications seem unduly pessimistic. The beatnik culture was a short-lived perturbation that was succeeded by a more energetic participation in the traditional rat race of the free-enterprise culture. Increasing affluence led to a desire for even more affluence.

In retrospect, Stent's analysis was perhaps unrealistically linear. He did not recognize that progress is a many-faceted thing. Progress in one area may create problems elsewhere. It was not the overall level of ease that was the important factor in society; it was the perceived differences in the level of success between one person and his or her neighbours. These inequalities are likely to be far more instrumental as a motivating factor than is the overall level of prosperity. Even without these inequalities, increasing peace and prosperity is a subtle thing. We have come to appreciate that technological progress has a serious downside. It often creates environmental problems that outweigh the benefits that the technology was designed to alleviate. If there are similar negative by-products of other forms of technical progress, then overcoming them will remain a constant stimulus to the human imagination. Stent's decadent Golden Age may never come.

Horgan sees a different type of future for science. Whereas Stent wonders whether the psychological motivation for science might wane, undermined by peace and security outside science, Horgan wonders whether all the answerable questions will dry up and science be undermined by decadence from within. Could all areas of fundamental inquiry soon reach frontiers of fascinating speculation that are not open to definite test by experiment or observation?

At first sight this seems very likely. Our own situation in the Universe and our technical capabilities have not been 'designed' with a view to the completion of *our* knowledge of the Universe. There is no reason to believe that the Universe exists for our convenience or amusement. There will be some limits to what we can do and know. If there are limits, and knowledge is cumulative, we can only be approaching them—there is no alternative. Eventually, we shall inevitably reach a state of knowledge that admits significant 'progress' only by drawing up plausible scenarios. No experiment will be able to distinguish them or exclude them decisively. This 'naïve ironic science', as Horgan dubs it, will provide interesting after-dinner conversation, and may even launch a thousand popular science books, but it will never help anyone to build a better machine or add to the canon of secure scientific knowledge. In some ways this future for the scientific enterprise is reminiscent of the fate of many of the creative arts. There, the 'ironic' label singles out the postmodernist attitude that there is no core of reader-independent truth at the root of the work. The text is what you find it to be. All texts possess multiple reader-dependent meanings, and the only 'true' meaning is the text itself. Literary criticism has entered a deconstructivist phase which maintains that any interpretation of a work is as valid as any other—

including that of its author.[13] Thus, Horgan's see those who work in fundamental physical sciences facing the future in which they have

> to pursue science in a speculative, postempirical mode that I call ironic science. Ironic science resembles literary criticism in that it offers points of view, opinions, which are, at best, interesting, which provoke further comment. But it does not converge on the truth. It cannot achieve empirically verifiable surprises that force scientists to make substantial revisions in their basic description of reality.[14]

Perhaps science faces such a subjective fate: one that many scientists would regard as a fate worse than the death of science. Regardless of the psychological question of whether such speculation has special attractions for successful empirical scientists at certain stages of their career, this is really a prediction about the nature of the Universe. It is an expectation that there is a limit to our observational handles upon the nature of things. There will be things that we cannot see, events that we cannot record, possibilities that we will not be able to rule out. When that happens all we can do is to paint pictures of possible scenarios that are consistent with what little we do know. But the gaps that remain in our knowledge will allow many different possibilities to exist. Whereas these lacunae form a small part of science today, their relative size may steadily grow. One day, our descendants may wake up to find that they may have grown to encompass the whole of the boundary between the known and the unknown.

In recent years the pronouncements and predictions of science have become increasingly bold and speculative. Scientists seem no longer content merely to describe what they have done or what Nature is like; they are keen to tell their audience what their discoveries *mean* for an ever-widening range of deep philosophical questions ('meaning-of-life issues'), and to speculate about future possibilities in ways that seem closer to the realm of science fiction than to science fact. Examples spring easily to mind: the quest to fabricate artificial forms of intelligence, the search for advanced extraterrestrial beings, the explanation of human feelings and emotions by adaptive evolution, the possibilities for reading the genetic code of life and redrafting crucial parts of its story to eradicate disease and extend the human lifespan far beyond its present length. Cosmologists tell us about the beginnings of our Universe (and others!) and prognosticate about the form of the ultimate laws of Nature, while others chart our eternal cosmic future. Each of these examples is a story in its own right, but one might ask whether the speculative reach of popular science is telling us something deeper about the nature of the subjects it expounds and the audience it caters for.

Some might see this penchant for the transcendental in the popularization of science as a substitute for the decline of traditional religions. Many see science as a source of transcendental ideas which take us beyond the humdrum alternatives of news about politics, scandal, economic issues, crime, and social fashions. The fascination with the occult, with astrology, and with other mystical yearnings to be at one with the Universe (witness the bizarre appearance of the so-called 'Natural Law Party', with attendant gobbledygook, in recent British and American elections). There seems to be a deep human desire for something larger than ourselves and for an understanding of the meaning of the Universe. Some writers have tried to latch on to this, quite deliberately. Paul Davies, for example, has claimed that science offers a surer road to God than does religion.[15] This is by no means a new claim. In 1932, the influential mathematician and physicist, Hermann Weyl considered this question in some detail, arguing that

> Many people find that modern science is far removed from God. I find, on the contrary, that it is much more difficult today for the knowing person to approach God from history, from the spiritual side of the world, and from morals; for there we encounter the sufferings and evil in the world which it is difficult to bring into harmony with an all-merciful and all-mighty God. In this domain we have evidently not yet succeeded in raising the veil with which our human nature covers the essence of things. But in our knowledge of physical nature we have penetrated so far that we can obtain a vision of the flawless harmony which is in conformity with sublime reason.[16]

Traditional science fiction has a much harder job staying in business than theology,[17] as science regularly uncovers possibilities more unusual than any fiction writer has yet imagined. It has used this to its advantage though, widening its scope and exploring psychological and non-technical problems in greater depth.

One might also wonder whether the market success of popular science has stimulated expositors to become increasingly speculative in their desire to attract readers. But there is a more straightforward possibility. Scientific disciplines have a 'filling factor' that is a measure of how completely they have uncovered what is currently within the reach of our experimental accuracy, computer technology, and human mathematical facility. As all the accessible results get swept up and explained in simple terms to outsiders, the only place left to go is to the speculative margins of the subject (and beyond). A proliferation of highly speculative extrapolations beyond what is currently known of a subject is a sign either that new observational facts are very difficult to uncover (as in the study of the Universe's distant past, for example), or that the branch of science in question has been so successful in uncovering what can be found within its domain that relatively little accessible information remains (as in experimental particle physics).

Progress and prejudice

An optimist is someone who thinks the future is uncertain.

ANONYMOUS

The assumption of constant progress is a relatively modern one.[18] It is a consequence of living long and living fast. Life in the past was slower; communication was harder; change was more difficult to promote; and far fewer people were able to bring it about. For most, there was little or no correlation between change and improvement; life was a treadmill, with little to gain and everything to lose.

In some cultures, progress could be hindered by deep-seated beliefs about the course and purpose of history. Many Eastern societies held fast to a tradition of cyclic recurrence by analogy with the seasonal variations and the cycle of birth and death witnessed in the natural world.[19] Christianity saw human history as a retreat from a paradise which would one day be re-established for the Elect. These are not views that sit easily with that of steady human progress over the course of history.

In medieval times philosophers and scientists spent more time looking backwards than looking forwards. The classical works of Aristotle were widely regarded as both necessary and sufficient for the understanding of all things. They provided an authority against which new ideas were tested and accommodated. Observation was not seen as the pre-eminent tool for sifting fact from fiction that it is today. Galileo could not convince the professor of philosophy at Pisa that the best way to judge his claim that Jupiter possessed moons was to look through his telescope and see. A by-product of this exaggerated respect for texts and authorities was the view that the Golden Age of insight and discovery lay in the past. The great philosophers had lived in ancient Greece: Plato and Aristotle were 'the giants on whose shoulders we stand'. We could not hope to surpass them.

The Renaissance relinquished this exaggerated respect for the past. The painters, sculptors, and scientists of the Renaissance showed that they could do better than their predecessors. The rebirth of confidence in human abilities that blossomed then instilled a feeling for progress and achievement that has continued until modern times.

The growth of applied science provided good measure of progress if any were sought. For example, the accuracy with which time could be kept was always a benchmark in seafaring countries because it determined how accurately longitude could be determined. During Newton's day huge sums of money were offered by the Admiralty in England as prizes for the designers of the most accurate timepieces to be used at sea for navigation.

The ancients also bequeathed to us a different attitude to the future. Aristotle, like many other thinkers who followed him, had laid great stress upon the place of 'purpose' in explaining how and why things happen. This seems clear when dealing with humans and animals, but becomes rather misleading when dealing with inanimate objects. Aristotle maintained that changes had purposes and goals in the future, called 'final causes', which revealed why they had occurred. This view became wedded to anthropocentric design arguments in the life sciences, which saw the structure of the living world as a product of design. The close match between the conditions required for particular living things to exist successfully and the structure of their environment was interpreted as evidence of Divine pre-programming.[20] One consequence of this teleological view is that the present state of the world comes to be regarded as the best possible in some sense. There need be no further progress towards some state of better adaptation between living things and their habitats. If you extol the wonders of the human eye as an optical instrument, then you cannot imagine progress or improvement without tacitly admitting it has imperfections.[21] Outside the life sciences, other, more subtle forms of this way of thinking existed. They appealed, not to the remarkable matches between aspects of the environment and the functioning of living things, but to the wonderful simplicity, universality, and appropriateness of the *laws* of Nature that Newton had revealed, which govern the structure of the Earth and the solar system.

An appreciation of change is most likely to come from the study of living things. But biology is not like astronomy: although change in living things is easy to see, it is difficult to understand: the past cannot easily be reconstructed, and there are no simple mathematical equations which predict the future. Life is too complicated. The central problem was to arrive at a convincing explanation of how living things came to be (rather than merely a 'Just So Story' of the sort that 'things are as they are because they were as they were'), and why they are seemingly tailor-made for their environments.

The first attempt to do this convincingly was made by the French zoologist Jean Baptiste de Lamarck (1744–1829). Lamarck appreciated the fact that organisms are always well adapted to their environments. But he saw a big problem. Environments change. So, organisms had better change as well if they are going to stay adapted to their circumstances. Lamarck's theory was that organisms learn new behaviours, or develop new structures, in response to environmental variations. These changes are gradually reinforced by repeated successful application. Organisms take their marching orders from the environment in some way. As trees grow taller, so giraffes will slowly develop longer legs or necks so that they can carry on feeding off the leaves. The result was a continuing harmony between the structure of organisms and their needs. Underlying this whole picture was a belief that living things tend to evolve towards the most harmonious and perfect forms. And there they stay. The major hole in Lamarck's theory was, of course, the lack of any mechanism by which information about environmental change could be conveyed to organisms so that their bodies 'knew' that they must change.

In the middle of the nineteenth century, Darwin and Wallace independently proposed a theory of evolution by natural selection that was radically different from Lamarck's. Darwin realized that the environment was an extremely complicated cocktail of competing influences and changes. There is no reason why its vagaries should be linked to the changes within living things at all. Something far simpler would do. He recognized that when changes occurred within an environment, all that happened was that some organisms found themselves able to cope with the new environment, while others did not. The former survived with a higher probability of passing on the attributes that enabled them to survive, while the others did not. In this way, those features which favoured survival in a particular environment (and could be inherited) were in the long run, preferentially passed on to future generations. This process is 'natural selection'. It does not guarantee that the next generation will be well adapted. If the environment changes suddenly, then the good adaptations of the past might even become liabilities. If the environment changes too dramatically, then an organism may not be able to adapt fast enough to survive and will become extinct.

An environment presents challenging problems for organisms, and the only resources available for their solution are to be found in the variations that occur in a breeding population. If the environment changes over a long period, then the preferential survival of those members of a species best able to cope with the environmental changes will result in a gradual change in the species. Successful adaptations will tend to survive, but there is no reason why they will in any sense be the best possible. In practice, this process of evolutionary adaptation can be very complicated because an organism's environment contains other organisms and is itself changed by the organisms present. It is therefore more accurate to talk of the coevolution of different organisms together with their environments, rather than of the evolution of a single organism or species.

Unlike Lamarck, Darwin saw breeding organisms as producing a variety of traits, at random, before there was any need for them. No unseen hand exists which generates only those variations that would be required to meet the pressing requirements of the immediate future. The useful ones are selected because they increase fecundity in the long run.

There is much more to be said about the process of natural selection, but for our story it is enough to draw out one central lesson. Natural selection killed the idea that the world is a finished product arrived at by design. Design is unnecessary. A finished world is unstable and would require constant readjustment to maintain its state of perfect adaptation in the face of changing environments. To keep up with all the natural changes would require a complicated reciprocal process—and that process would be natural selection.

Nature is not like a clockwork mechanism. An unfinished watch does not work. The world has a future that differs from the present. If we wish, we can call the difference between the future and the present 'progress', so long as we appreciate that it might well turn out to be negative in certain respects, even if it is positive in others.

After Darwin, there were many attempts to extend the idea of evolution into social affairs and explain everything and anything by the same principle of the 'survival of the fittest'. Few of these speculations were well founded but they gave rise to a particular concept of progress and a direction of change.[22] We shall have more to say about this in Chapter 5, when we look at the progress of technological capability.

We can see that evolution did away with the idea that the living world is a finished product. This opens the door to ideas of progress (and regress) and to speculations about what the world might be like in the future. These ideas come more naturally to life scientists. Physical scientists who study the mathematical laws of Nature lay much emphasis upon the unchanging character of those laws. Before the twentieth century, the most successful applications of those laws were to the motions of the Moon and the planets. The changes seen in the astronomical realm were slower, simpler, and more predictable than those in the living world. Not until the twentieth century would astronomers have to come to terms with radical new theories about the origin and evolution of stars and galaxies, and the discovery of the expansion of the Universe.

Newton's discoveries had been so impressive for nearly two hundred years that they had the hallmark of being the last word. No refinements of his laws had been suggested. His law of gravitation had successfully explained every astronomical observation (with the tiny exception of a wobble in the orbit of the planet Mercury around the Sun). In fact, during his own lifetime the success of his mechanics had led to speculations that his approach might provide a panacea for the investigation of all questions. The impressive completeness of Newton's *Principia* (1687) and the deductive power of his mathematics led to a bandwagon effect with thinkers of all shades aping the Newtonian method. There were books on Newtonian models of government and social etiquette, and Newtonian methods for children and 'ladies'.[23] Nothing was imagined to be beyond the scope of the Newtonian approach. Nor was Newton himself entirely divorced from this enthusiasm. His later work on alchemy and biblical criticism reveals a deep-rooted belief in his ability to unveil all mysteries for the human race. Having first revealed the truth about God's design of the physical world, he seems to have seen himself as having a similar commission to fulfil in the realm of the spiritual and the mystical.[24] Newton is a deeply paradoxical figure when viewed through the lens of modern scientific attitudes. A mathematical genius who possessed the most penetrating physical intuition of any recorded scientist, he nevertheless had one foot in the Middle Ages and displayed a magician's belief in his ability to solve all problems and overcome all barriers. His achievements must have made his contemporaries believe that the end of the seventeenth century was indeed the completion of science.

The big idea of unlimited knowledge

Definition: *Science is systematised positive knowledge, or what has been taken as such at different ages and in different places.*

Theorem: *The acquisition and systemisation of positive knowledge are the only human activities which are truly cumulative and progressive.*

Corollary: *The history of science is the only history which can illustrate the progress of mankind. In fact, progress has no definite and unquestionable meaning in other fields than the field of science.*

GEORGE SARTON[25]

Nineteenth-century commentators displayed almost every possible attitude towards the future of science. There were those who thought that the completion of science was possible in principle, but not in practice, and there were others who sought to distinguish carefully between the certainties of different types of knowledge. In this latter respect, the most significant new turn was the distinction made between the world as it really is and our perception and apprehension of it. This distinction, made carefully by the German philosopher Immanuel Kant in the eighteenth century, argued that our apprehension of the world was always processed through the mental concepts that the brain provided. Something was always left out or distorted in that process. We cannot have access to the raw unexpurgated truth about things. There must always be a gap between reality and our knowledge of it. Thus, Kant revealed, there is a fundamental limit to our knowledge of things: an unbreachable gap between what is and what we can know about it.

While this gap is undeniable, there is still room to argue about how big it is. If the distortion is very small we might be able to ignore it with impunity. Alternatively, it may be that our mental processes are specially conformed to receiving certain sorts of information about the world and so, when considering those aspects of things, the distortion is minimal, or even zero. What we have learnt about natural selection gives some credence to the latter view because we now know, as Kant did not, that the categories of thought that we use to make sense of the world are the results of a process of natural selection. They have presumably been selected for their success in giving an accurate representation of those parts of reality which are important for the survival of organisms. This could explain why your picture and impression of the world seems to be so similar to mine.[26] One must be a little wary of this as a catch-all defence against the distortion of reality by our categories of understanding. Not all those categories need be direct consequences of evolution. If they are by-products of natural selection for other abilities and functions then they need not be optimal at all. It is the entire collection of human abilities that will determine survival.

In one area of human inquiry there had long existed a quiet confidence in our ability to fathom something of the ultimate truth about the Universe. And if this success was possible in one area of inquiry, it was believed, then why not in others too? The source of this confidence lay in the age-old study of geometry that Euclid and the ancient Greeks had placed upon a firm logical foundation.

The great success of Euclidean geometry had done more than help architects and cartographers. It had established a style of reasoning, wherein truths were deduced by the application of definite rules of reasoning from a collection of self-evident axioms. Theology and philosophy had aped this 'axiomatic method', and most forms of philosophical argument followed its general pattern. In extreme cases, as in the works of the Dutch philosopher Spinoza, philosophical propositions were even laid out like the definitions, axioms, theorems, and proofs in Euclid's works.[27]

The most important consequence of the success of Euclidean geometry was that it was believed to describe how the world was. It was neither an approximation nor a human construct. It was part of the absolute truth about things. Thus our understanding of it was very encouraging. It underwrote confidence in human ability to fathom the absolute truth about the world. If a theologian was criticized for asking questions about the Divine Nature on the ground that such absolute truths are beyond our reach, he could point to Euclidean geometry as evidence that some of these truths are accessible to us—and if some, why not others?

This confidence was suddenly undermined. Mathematicians discovered that Euclid's geometry of flat surfaces was not the one and only logically consistent geometry. There exist other, non-Euclidean, geometries that described the logical interrelationships between points and lines on curved surfaces (see Fig. 2.1). Such geometries are not merely of academic interest. Indeed, one of them describes the geometry on the Earth's surface over large distances. Euclid's

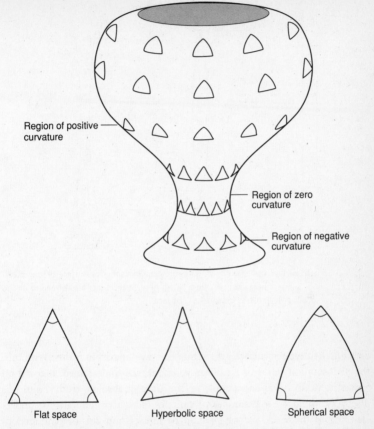

Fig. 2.1 A vase possesses regions which have spherical, hyperbolic, and flat (Euclidean) geometry. These three geometries are defined by the sum of the three interior angles of a triangle formed by drawing the shortest distances between three points on the surface. This sum exceeds 180 degrees in a spherical space, is less than 180 degrees for a hyperbolic space, and equals 180 degrees for a flat Euclidean space, as shown.

geometry of flat surfaces happens to be a very good approximation locally only because the Earth is so large that its curvature will not be noticed when surveying small distances. Thus, a stonemason can use Euclidean geometry, but an ocean-going yachtsman cannot.

This simple mathematical discovery revealed Euclidean geometry to be but one of many possible logically self-consistent systems of geometry. None had the status of absolute truth. Each was appropriate for describing measurements

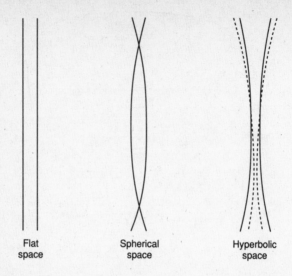

Flat
space

Spherical
space

Hyperbolic
space

Fig. 2.2 On a flat surface, only parallel lines never meet. On a spherical surface all lines meet. On a hyperbolic surface there are many lines which never meet. Lines are defined to be the shortest distances between two points on these surfaces.

on a different type of surface, which may or may not exist in reality. With this, the philosophical status of Euclidean geometry was undermined. It could no longer be exhibited as an example of our grasp of absolute truth. From this discovery would grow a number of varieties of relativism about our under-standing of the world.[28] There would be talk of non-Euclidean models of government, of economics and of anthropology. 'Non-Euclidean' became a byword for non-absolute knowledge. It also served to illustrate most vividly the gap between mathematics and the natural world. There were mathematical systems that described aspects of Nature, but there were others that did not. Later, mathematicians would use these discoveries in geometry to reveal that there were other logics as well. As a result, even the concept of truth was not absolute. What is false in one logical system can be true in another. In Euclid's geometry of flat surfaces parallel lines never meet, but on curved surfaces this is no longer true (Fig. 2.2).

These discoveries revealed the difference between mathematics and science. Mathematics was something bigger than science, requiring only consistency to be valid. It contained all possible patterns of logic. Some of those patterns were followed by parts of Nature; others were not. Mathematics was open-ended, incompleteable, infinite; the Universe might not be.

Negativism

The Titanic *sails at dawn.*

BOB DYLAN[29]

Look back through the history books and you will find at any time a mixture of optimists and pessimists regarding the likely progress of science.[30] During the nineteenth century the pessimists' team was organized into a philosophical movement that was known at the time, rather inappropriately, as *positivism*.[31] We shall refer to it, more appropriately, as *negativism*.

Positivism was promulgated by the French philosopher Auguste Comte (1798–1857), and was subsequently taken up by the influential scientist and philosopher of science Ernst Mach. Mach's views about motion made a deep impression upon Einstein, and influenced the thinking that culminated in his theories of special and general relativity. But Mach was an extremely con-servative philosopher. Like Comte he insisted upon limiting the arena of reliable human knowledge to those phenomena about which we could have direct sense perceptions. This had an unfortunate effect. Instead of simply making scientists far more demanding of evidence, and far more critical in their evaluation of their theories, this philosophy appears to have discouraged scientists from investigating many areas where new discoveries would have been possible.

Comte gave a number of specific examples of problems that he believed that it would be impossible for us to answer.[32] He saw the evolution of human thinking as a process that must pass through three stages. The first two stages of this trinity, the theological and metaphysical, were signs of immaturity and merely precursors of the third and most desirable *positive* stage. In the 'theological' stage, the human mind is still at the necessary starting point of intelligent inquiry, but at this stage

> the human mind directs its researches mainly toward the inner nature of beings, and toward the first and final causes of all the phenomena that it observes—in a word, toward absolute knowledge. It therefore represents these phenomena as being produced by the direct and continuous action of more or less numerous supernatural agents, whose arbitrary intervention explains all the apparent anomalies of the universe . . . [the theological stage] arrived at its highest form of perfection when it substituted the providential action of a single being for the varied play of the numerous independent gods which had been imagined by the primitive mind.[33]

During the next, or 'metaphysical' stage, he sees things improving, but only a little, because

> In the metaphysical state, which is in reality only a simple general modification of the first state, the supernatural agents are replaced by abstract forces, real entities

or personified abstractions, inherent in the different beings of the world. These entities are looked upon as capable of giving rise by themselves to all the phenomena observed, each phenomenon being explained by assigning it to its corresponding entity . . . The last stage of the metaphysical system consisted in replacing the different special entities by the idea of a single great general entity—nature—looked upon as the sole source of all phenomena.[34]

Finally, in the third 'positive' stage the mind has given up hankering after explanations for the unexplainable and the vain quest for answers to ultimate questions. Now, having matured,

the human mind, recognizing the impossibility of obtaining absolute truth, gives up the search after the origin and hidden causes of the universe and a knowledge of the final causes of phenomena. It endeavours now only to discover, by a well-combined use of reasoning and observation, the actual laws of phenomena—that is to say, their invariable relations of succession and likeness. The explanation of facts, thus reduced to its real terms, consists henceforth only in the connection established between different particular phenomena and some general facts, the number of which the progress of science tends more and more to diminish.[35]

By analogy with the ultimate results of the first two stages of reasoning, this third stage has an ideal goal

toward which it constantly tends, although in all probability it will never attain such a stage, [which] would be reached if we could look upon all the different phenomena, observable as so many particular cases of a single general fact, such as that of gravitation, for example . . . the fundamental character of the positive philosophy is to consider all phenomena as subject to invariable natural laws. The exact discovery of these laws and their reduction to the least possible number constitute the goal of all our efforts; for we regard the search after what are called causes, whether first or final, as absolutely inaccessible and unmeaning.[36]

Comte urged scientists to content themselves with working models of Nature, like Newton's law of gravitation, and not to seek the cause of gravity or source of heat because he saw these deeper causes as unknowable. One can see already what an unsatisfactory philosophy of science this is likely to prove. While it might be true that we are unable to obtain complete or ultimate understanding of the nature of a force like gravity, there is no telling how far from that nirvana we are at the moment. Further investigation might well deepen our understanding by relating gravity to other forces, or to other aspects of the structure of the Universe. Although he saw a unification of scientific laws into a single law of Nature as the ultimate goal of human inquiry, he did not believe that this ultimate knowledge was humanly attainable, for,

It is my deep personal conviction that these attempts at the universal explanation of all phenomena by a single law are highly chimeral, even when they are made by the most competent minds. I believe that the resources of the human mind are too feeble, and the universe is too complicated, to admit of our ever attaining such scientific perfection; . . . It seems to me that we could hope to arrive at it only by connecting all natural phenomena with the most general positive law with which we are acquainted—the law of gravitation—which already links all astronomical phenomena to some of the phenomena of terrestrial physics . . . While trying to diminish as far as possible the number of general laws necessary for the positive explanation of general phenomena . . . we shall think it rash ever to hope, even in the most distant future, to reduce those laws rigorously to a single one.[37]

Comte picked on four specific areas where he believed that scientific inquiry was limited by an inability to obtain 'positive' knowledge—that is, direct sense data. In the realm of astronomy he discounted the possibility of positive knowledge of the stars. He thought (wrongly) that there was no way in which we could ascertain their chemical composition and (rightly, so far) that there were many unseen stars that optical observations cannot detect (what astronomers would now call 'dark matter'). Although he regarded astronomy as the pinnacle of positive science because it was free of direct metaphysical and theological contamination, his views were strangely geocentric. He ridiculed the discovery of Neptune as 'a so-called discovery, which, even supposing it genuine, could have no real interest except for the inhabitants of Uranus', and thought astronomy worthwhile only in respect of studying how things influence the Earth, arguing that

> When all heavenly bodies were supposed to be connected with the earth, or rather subordinate to it, it was reasonable that none should be neglected. But now that the earth's motion is known to us, it is not necessary to study the fixed stars, except so far as they are required for purposes of terrestrial observation . . . Even supposing it possible to extend our investigations to other [solar] systems, it would be undesirable to do so. We know now that such investigations can lead to no useful result: they cannot affect our views of terrestrial phenomena, which alone are worthy of human attention[38]

He discounted biology and chemistry as subjects in which mathematics could be usefully employed, rejected attempts to seek a deeper understanding of heat, light, and magnetism, and dismissed the use of statistical reasoning as irrational. He opposed the use of the concept of 'atoms' as building blocks of matter, believing that the 'ultimate structure of bodies must always transcend our knowledge'. Historians still debate the suggestion that Comte's views were partly responsible for the subsequent decline in French science.[39]

One of the curiosities of Comte's claims about the evolution of human inquiry through the three stages of theological, metaphysical, and positive knowledge is the way in which they look like the reverse of the trend that Horgan predicts. As the empirical content of a direction of inquiry is exhausted it will enter a stage of metaphysical analysis—'what does this knowledge mean?', 'could the world have been different?', 'why are things like this?', and so forth— to be followed by one of theological analysis—'why is there something rather than nothing?', 'is what we know compatible or incompatible with the existence of God?', 'what does our knowledge tell us about the origin, purpose, and ultimate fate of life in the Universe', and so on.

In both analyses there are great simplifications. Not only are all scientists treated as though they were a single individual ('science'), but individual scientists are treated as though they pursue one and only one activity. In reality, modern scientists who have an interest in ultimate cosmological questions, for example, usually have many other research interests with direct links to observational astronomy or to the study of mathematical structures.

Some nineteenth-century ideas of the impossible

> *Given for one instant a mind which could comprehend all the forces by which nature is animated and the respective situation of the beings who compose it—a mind sufficiently vast to submit these data to analysis—it would embrace in the same formula the movements of the greatest bodies of the universe and those of the lightest atom; for it, nothing would be uncertain and the future, as the past, would be present to its eyes.*
>
> LAPLACE[40]

When mathematics was used to describe the patterns that Nature wove in space and time the results were often spectacularly successful: none more so than in the description of the celestial motions. Newton's laws were held up as the paragon of scientific determinism. If you knew the present they would enable you to reconstruct the past and predict the future. This success led two great scientists to speculate about what Newton's laws might allow us to know if we had superhuman capabilities. These speculations are interesting because they create a picture of what it might be like to have unlimited knowledge. They do this by considering a limiting process that starts with ourselves and produces an omniscient being simply by magnifying our own abilities. This leads to the idea of an omniscient being who needs to be only quantitatively different to ourselves (rather than qualitatively different). Let us see what they had in mind.

Our two scientists, Laplace and Leibniz, both saw that the laws of Nature which Newton had discovered created a situation in which the future might be completely known by a mind large enough to know the present state of the Universe completely and to carry out the calculations required to predict its future state. Although it was granted that we are far from being able to achieve this level of knowledge and computational prowess, these determinists saw the difference between us and such a supermind as one of degree rather than of kind. What is interesting about their speculative conceptions is the fact that they open the door on the idea of complete knowledge. This optimism springs, not from hopes about future progress, but simply from a fuller application of the knowledge that they already had. Leibniz extended this optimism into an even wider domain. He conceived of a symbolic manipulation procedure that could be programmed with the laws of logic. It would be able to decide whether any statement was a true or false consequence of the logical axioms. Somewhat optimistically, he imagined that this formal procedure would allow all sorts of human disputes to be resolved logically. For instance, religious truths could be deduced rigorously after the manner of mathematical proofs, so putting an end to countless theological disputes. Again, one sees the concept of limitless (although not complete) knowledge. The concept requires just an extension of mundane abilities that we already possess rather than some great qualitative amplification of human abilities. The implication of these conceptions was that all questions might be answered by a systematic approach to them. There was certainly no acceptance of limits to the scientific enterprise which could not be overcome by progressive enlargement of our abilities. Today, the response to the super-being conceived of by Laplace is rather different. We know that to locate precisely every particle of matter in the Universe, together with its state of motion, is not merely difficult: it is impossible in principle. The quantum picture of matter teaches us that there is a fundamental limit to our ability to determine simultaneously the location and motion of any particle of matter. This might not be so bad if it were the case that small errors did not really matter. But, on the contrary, we have become progressively aware of the fact that it is typical of natural systems that they exhibit an extraordinary sensitivity to their precise position and motion. Thus, if we slightly alter the movement of a molecule of air, it will subsequently diverse very rapidly from where it would have been had it not been disturbed. This sensitivity has become known as 'chaos'. It means that Laplace's superbeing could not know the locations and motions of all the components of the world with sufficient accuracy to predict even the weather with 100 per cent accuracy if he obeys the laws of physics. This last caveat is important, because Laplace was not talking about *our* ability to predict the future of the heavenly bodies. He was talking about 'a mind . . . [of which] the human intellect offers, [only] in the perfection to which it has brought astronomy, a faint idea of what such a mind would be.'

While Comte supplied general scepticism about the growth of human knowledge, and Laplace typifies a certain over-confidence in determinism, there was a third strand to the nineteenth-century limits-of-science debate that was in some ways more interesting because it supplied a list of insoluble problems.

As one might expect, the limits of science were increasingly discussed near the end of the nineteenth century. The most influential event was the argument between two German scientists who were also influential in communicating scientific issues to a wider public. Emil du Bois-Reymond was a physiologist, philosopher, and historian of science; his opponent Ernst Haeckel was a zoologist with strong humanist and Monist philosophical leanings.

In 1880, Du Bois-Reymond published the text of two influential public lectures on the limits of science,[41] delivered in 1872 and 1880, the latter on the occasion of the Leibniz celebration of the Prussian Academy of Sciences in Berlin in July of that year. He believed there were definite limits to science because there were limits to the application of mechanical explanations and methods of experiment. For Du Bois-Reymond, natural science was about the motions of atoms: just 'the resolution of natural processes into the mechanics of atoms'. His claims were more challenging because he attempted to identify the insoluble problems: 'The Seven Riddles of the Universe', as he called them. Taking up the concept of Laplace's super-being, Du Bois-Reymond considers an image of a great mathematical theory of everything which could be used to predict the future course of the Universe from its present state;

> we may conceive of a degree of natural science wherein the whole process of the universe might be represented by one mathematical formula, by one infinite system of differential equations, which would give the locations, the direction of movement, and the velocity, of each atom in the universe at each instant.

By running it backwards it could tell us how the world began, for

> if in his universal formula he set down $t = -\infty$, he could discover the mysterious primeval condition of all things.

Alternately, if we ran its predictions forward into the far future we could discover if the Universe was steadily winding down, like one of Carnot's heat engines,[42] and approaching a state of complete equilibrium and 'Heat Death'; for,

> Suppose he lets $t[ime]$ grow *ad infinitum* in the positive sense, then he could tell whether Carnot's theorem threatens the universe with icy immobility in finite or only in infinite time.

Having introduced the concept of unlimited knowledge, Du Bois-Reymond moves on to consider its limits when faced with limitations of the human senses. He believes that the differences between Laplace's omniscient being and the human mind, although vast, are only ones of degree; indeed:

> We resemble this mind, inasmuch as we conceive of it. We might even ask whether a mind like that of Newton does not differ less from the mind imagined by Laplace, than the mind of an Australian or of a Fuegian savage differs from the mind of Newton.

Our limitations are clear: we are never going to be able to get all the facts that we need to put into the universal formula. Although in principle we might use the universal formula to reconstruct the past and predict the future, in practice we cannot:

> the impossibility of stating and integrating the differential equations of the universal formula, and of discussing the result, is not fundamental, but rests on the impossibility of getting at the necessary determining facts, and, even where this is possible, of mastering their boundless extension, multiplicity and complexity.

After these general issues, Du Bois-Reymond turns to his seven insoluble problems. They fall into two categories. The first group of four consists of several difficult, but potentially soluble, problems: *the origin of life, the origins of language and human reason*, and *the evolutionary adaptiveness of organisms*; the second consists of two problems which he regards as insoluble in principle, and a third which may turn out to be of a similar nature. The first group of problems were well chosen. Their importance remains primary even today. They are problems about which we now know a considerable amount, but none of them could be said to be 'solved'. We know most of the underlying pieces of the puzzle from which a solution will ultimately emerge in each case, but not all those pieces are yet in place. There is no reason, however, to believe these problems contain any special degree of insolubility over and above any complicated scientific problem. The second group of problems is altogether different, and it is intriguing to dwell on them a little longer in order to compare Du Bois-Reymond's thinking with our own, as well as with that of his contemporaries of a more optimistic persuasion. His choices are tantalizingly close to the most common themes of contemporary popular science writing. Here they are:

Insoluble Problem Number 1: the origin of natural forces and the nature of matter

Du Bois-Reymond questions whether we can ever do better than represent matter by some conceptual model: 'atoms' in this case were not atoms in the modern chemical and physical sense; merely the smallest elements of matter). If we imagine these building blocks to be infinitesimally small, then this may be nothing more than 'a useful fiction in mathematics' or a 'philosophical atom'. He also worries about the old problem of how a force like gravity can act across empty space, and argues that we have developed our concepts of force and matter by extrapolating our limited sensual experiences of them. By extrapolating so far beyond the realm of sense data we have unreliable knowledge, and we are fooling ourselves if we count all these extrapolations as solid advances in knowledge. In reality, they are merely speculations. Du Bois-Reymond believes this problem to be insurmountable, even for Laplace's supermind:

> no one . . . can fail to acknowledge the transcendental nature of the obstacles that face us here. However we try to evade them, we ever meet them in one form or another. From whatever side we approach them, or under whatever cover, they are ever found invincible . . . For even the mind imagined by Laplace, exalted as it would be high above our own, would in this matter be possessed of no keener insight than ourselves, and hence we despairingly recognise here one of the limitations of our understanding.

What Du Bois-Reymond claims is that we can never know the ultimate elementary particles and forces of Nature. His argument is a pragmatic one. In order to explain these things properly we need to have the full facts about their nature. Those facts are ultimately hidden from us because they require us to be able to extrapolate down to infinitesimally small sizes and know all the forces that act there. This we cannot do.

Insoluble Problem Number 2: the origin and nature of consciousness and sensation

The second of Du Bois-Reymond's insolubilia is that of consciousness. He sees a twofold limitation here. The first is the problem of explaining what it is; the other arises because its existence leads to a breakdown in predictability for Laplace's demon. In the evolution of life on Earth, consciousness is

> something new and extraordinary; something incomprehensible, again, as was the case with the essence of matter and force. The thread of intelligence, which stretches back into negatively-infinite time, is broken, and our natural science comes to a chasm across which no bridge, over which no opinion can carry us: we are here at the other limit of understanding.

Du Bois-Reymond tries to imagine how we might approach the problems of sensation and mental activity as we do the problems of celestial mechanics, noting the position and velocity of motion for every particle, and then using Newton's laws to predict their future course. Applying such a method to a description of the brain, we might link the occurrence of certain mental phenomena to specific muscle responses. It certainly would be

> a great triumph of human knowledge if we were able to say that, on occasion of a given mental phenomenon, a certain definite motion of definite atoms would occur in certain definite ganglia and nerves. It would be profoundly interesting if we could thus, with the mind's eye, note the play of the brain-mechanism, in working out a problem in arithmetic, after the manner of a calculating machine; or, even if we could say what play of the carbon, hydrogen, nitrogen, oxygen, phosphorus, and other atoms corresponds to the pleasure we experience on hearing musical sounds; what whirl of such atoms answers to the climax of sensual enjoyment; and what molecular storm to the raging pain we feel when the trigeminus nerve is misused [and we have a headache].

But even if we had this type of knowledge of cause and effect in the human mind, Du Bois-Reymond argues that it would not help us understand *qualities* of sensual experience:

> What conceivable connection subsists between definite movements of definite atoms in my brain, on the one hand, and on the other hand such . . . undeniable facts as these: 'I feel pain, or pleasure; I experience a sweet taste, or smell a rose, or hear an organ, or see something red,' and the immediately-consequent certainty, 'Therefore I exist?' it is absolutely and for ever inconceivable that a number of carbon, hydrogen, nitrogen, oxygen, etc., atoms should not be indifferent to them.

Insoluble Problem Number 3: The problem of free will

Du Bois-Reymond finds the existence of our free will totally perplexing. It seems irreconcilable with a mechanical view of the universe. But he is less certain about how to characterize the problem. The extent to which it merits being classed as insoluble would ultimately be determined by the extent to which it falls within the bounds of the problem of consciousness.

Du Bois-Reymond's arguments excited considerable debate, not least because of his powerful position within the scientific establishment. The sharpest riposte came from the pen of the zoologist Ernst Haeckel in a widely read book, *The Riddle of the Universe*.[43] Clearly, Haeckel disliked Du Bois-Reymond almost as much as he disliked his views about the limits of science, calling himself one of the 'few who had sufficient scientific knowledge and moral courage to oppose the dogmatism of the all-powerful secretary and dictator of the Berlin Academy of Sciences'. Haeckel regarded the biological puzzles—the origins of life and language, and adaptiveness—as soluble problems that would be solved by application of the theory of natural selection. The problem of free will he viewed as a pseudo-problem because there was no evidence that it was anything other than a pure illusion, arguing that free will 'is a pure dogma [resting] on mere illusion and in reality does not exist at all'. Finally, the problems of substance, motion, and force he saw, in each case, as a confusion of two problems. The first was a philosophical rather than a scientific problem; the second, scientific part was, he believed solved by the laws of conservation of mass and of energy. His conclusion is that

> The number of world riddles has been continually diminishing in the course of the nineteenth century . . . Only one comprehensive riddle of the universe now remains—the problem of substance . . . [but today] we have the great, comprehensive 'law of substance', the fundamental law of the constancy of matter and force. The fact that substance is everywhere subject to eternal movement and transformation gives it the character also of the universal law of evolution. As this

supreme law has been firmly established, and all others are subordinate to it, we arrive at a conviction of the universal unity of nature and the eternal validity of its laws. From the gloomy *problem* of substance we have evolved the clear *law* of substance.[44]

Haeckel was in his own way as misguided as Du Bois-Reymond. He thought that science was fast approaching a state in which all the major problems would be solved. All that would be left were linguistic and philosophical questions about the meaning of those solutions. By contrast, Du Bois-Reymond thought that science was fast approaching an end of a different sort: an encounter with fundamental limits. In some sense their views are actually rather close. They both saw science as nearing the end of the road. Du Bois-Reymond believed that fundamental human limitations were responsible; Haeckel believed that the end was nigh because we would soon know all that constitutes scientific knowledge.

During the period from 1870 to 1905 Haeckel's optimistic view was widely shared. The American philosopher of science Charles Sanders Peirce[45] advocated a theory of truth which defined 'truth' to be the culmination of scientific investigation;[46] and, the great goals of scientific investigation all but achieved, many physicists believed that their subject was nearing completion. The young Max Planck recalled how, in 1875, as a young student he was steered towards the biological sciences by his mentors on the grounds that all the important problems of physics were already solved:

> As I was beginning to study physics and sought advice regarding the . . . prospects of my studies from my eminent teacher Phillip von Jolly, he depicted physics as a highly developed and virtually full-grown science, which—since the discovery of the principle of the conservation of energy had in certain sense put the keystone in place—would soon assume its stable form. Perhaps in this or that corner there would still be some minor detail to check out and coordinate, but the system as a whole stood relatively secure, and theoretical physics was fast approaching that degree of completeness which geometry, for example, had already achieved for hundreds of years.[47]

On the other side of the Atlantic the same sentiments were being voiced. The leading American physicist and future Nobel physics Laureate Albert Michelson claimed in a public lecture at the University of Chicago in 1894 that

> The most important fundamental laws and facts of physical science have all been discovered and these are now so firmly established that the possibility of their ever being supplanted in consequence of new discoveries is exceedingly remote, nevertheless, it has been found that there are apparent exceptions to most of these laws, and this is particularly true when the observations are pushed to the limit . . . our future discoveries must be looked for in the sixth place of decimals. It follows that every means which facilitates accuracy in measurement is a possible factor in a future discovery.[48]

The last quarter of the nineteenth century was a time when physicists liked to congratulate themselves on their past successes. Their enterprise was well advanced; all the great principles seemed to have been found. The conservation of energy, the laws of motion, gravity, electricity and magnetism, and thermo-dynamics seemed to be able to deal with anything that confronted them. The scepticism of philosophers was largely aimed at the human sciences, which were not well advanced, or at matters of such a basic nature, like the origin of matter, that they could justifiably be shifted into that realm of unanswerable problems that can be helpfully labelled 'philosophical questions'.

In retrospect, we can see that this period did indeed bring to a close a chapter in the development of physics. What is often called 'classical' physics was drawing to a close. But, far from being the end of physics, it was not even the beginning of the end. The revolution began in 1905. Soon there would be developments which would bring new theories of quantum mechanics, relativity, atomic structure, and gravitation on to the scene. Curiously, none of them was triggered by some new measurement of natural phenomena at unprecedented accuracy finding a new and unsuspected new layer of un-explained detail. All the revolutions would begin from within the heart of what was known.

Summary

The reasonable man adapts himself to the world: the unreasonable one persists in trying to adapt the world to himself. Therefore all progress depends on the unreasonable man.

GEORGE BERNARD SHAW[49]

In this chapter we have widened the scope of our thinking about the impossible by looking at how it defines, not only the existence of science, but also (to some degree) its limits, and the different ways in which they can arise. The great acceleration in scientific progress means that if there are limits, then they are being approached.

We looked at two distinctive claims that science is fast coming to the end of a road (if not *the* road). Ironically, both result from the success of science. Gunther Stent looked to a loss of the basic motivation for technical innovation, brought about by the increase of leisure and the lack of challenge that life increasingly presents to those in the Western democracies. The journalist John Horgan sees a different endgame for fundamental science. As the means of testing ideas has lagged farther and farther behind our ability to proliferate them, so the frontiers of science have become increasingly focused upon speculative ideas far removed from things that we can ever observe or test. Science therefore runs the risk of going the way of so much of the humanities, slipping into a mire of relativism, where there can be nothing more than opinions. By contrast, there are those who see science as an evolving progressive enterprise. We looked at the nineteenth-century background to this view, and at the contrary view, that science faced insoluble problems, which both came to such prominence in the closing years of the last century. The pessimistic views that captured so much attention at that time are especially interesting because some of them are so specific: they identify actual problems which will not be solved. Problems of the origins of life, matter, consciousness, and free will were well chosen. They are likely to be with us for some considerable time in the future.

CHAPTER 3

Back to the future

You cannot fight against the future. Time is on our side.
WILLIAM EWART GLADSTONE[1]

What do *we* mean by the limits of science?

An unwillingness to admit the possibility that mankind can have any rivals in
intellectual power occurs as much amongst intellectual people as amongst others:
they have more to lose.

ALAN TURING[2]

The simplicity of the phrase 'the limits of science' is deceptive. We are familiar with the limitations of scientists; we are familiar with partial theories about how things work; and we are familiar with theories that are simply wrong. In each case, what might at first be casually referred to as a limit of science is really nothing of the sort. Perhaps, after all, there are no true limits to science at all.[3] Perhaps all boundaries are illusory, whether erected by ourselves through our lack of information about the nature of things, or by the choice of an oversimplified (or even an over-complicated) model of reality? This is an issue that must be taken seriously. All our attempts to describe the workings of Nature and to predict or control future events are based upon a scientific method that builds up a 'model' of how some aspect of Nature operates. The more observations we make, the more completely and accurately this representation of Nature can be checked and extended. Our models of Nature are invariably mathematical in character. This is not as narrow as it first sounds. Although the outsider sees mathematics as a coldly analytical way of looking at the world, it is something deeper than this: something that is closely linked to other human pictures of the world. At root, mathematics is the name we give to the collection of all possible patterns and interrelationships. Some of those patterns are between shapes, others are in sequences of numbers, while others are more abstract relationships between structures. The essence of mathematics lies in the relationships between quantities and qualities. Thus it is the relationships between numbers, not the numbers themselves, that form the focus of interest for modern mathematicians. Accordingly, the subject abounds with terms like 'transformations', 'symmetries', 'programs', 'operations', and 'sequences' which describe relationships between things.

As soon as some aspect of the world is described by a model, say a system of mathematical rules, we are faced with some deep questions:

- Is the gap between reality and the mathematical description of it a harmless one?
- Does the use of a mathematical model introduce any limitation upon what we can deduce from the model?
- How can we distinguish limits imposed by our choice of model from limits that would be imposed by any (or no) choice of model to codify our observations of Nature?

At first we might think that using a computer to predict how some complicated natural phenomenon occurs is two steps removed from reality. However, we see that we are always limited in a very similar way by the human mind, which shares many of the properties of sophisticated computers. Any limitations on the scope of computing systems might well turn out to limit the power of human thought. In recent years there has been an upsurge of interest in the problem of human consciousness as scientists from a range of different disciplines try to put their finger on what it is. While some are confident that the human mind differs from a computer only in power and compactness, others have argued that it is qualitatively different. The most vocal supporter of this view is Roger Penrose,[4] who claims that the performance of feats of mathematical intuition catches the brain in the throes of doing something which no algorithmic computer can imitate.[5]

Before we begin to look at a modern view of science and its possible limits, it is interesting to gain some perspective by looking again at some predictions from the past. What sort of attitude has been taken to human progress in the last few centuries? Has there been outrageous overconfidence in the scope of human capabilities? Have past thinkers been too pessimistic in their expectations, or merely so lacking in the right sort of imagination that they appear to have had no conception of where things were leading? The most illuminating statements on this subject often come after periods of great success in some branch of science.

Possible futures

Millions long for immortality who do not know what to do with themselves on a rainy Sunday afternoon.

SUSAN ERTZ[6]

We are not very good at predicting futures. The bookmakers count upon it. Astrologers prove it. Suppose we were to wake up Rip-van-Winkle-style, thousands or millions of years in the future. What would the state of human knowledge be? How far would science have advanced? Would it be complete in any sense, perhaps because all accessible truths would have been found? Would some fundamental lines of inquiry have been finished? Would successors always spring up to take their places? We would be brave indeed to foretell the future. We shall try to do something that is rather easier: to outline some possible futures for the development of human knowledge about the Universe. But before exploring some plausible future scenarios, we should give some words of warning about simplistic notions of progress and sketch a new picture of scientific progress.

It is easy to fall into the trap of thinking that scientific progress is entirely cumulative: an inexorable accumulation of facts. But it's not really like that at all. Science does not only progress by making new discoveries. Sometimes it advances by showing that existing ideas are wrong, or that past measurements were biased in some way. The general trend may be advancing, just like the flow of a river, but like the motion of a leaf on the water surface, its path may meander back and forth.

There have been many pictures of how science grows. Four are especially interesting, not least because of the eloquence of their espousal. The first is the image of the tide, put forward by the French physicist Pierre Duhem:

> Scientific progress has often been compared to a mounting tide; applied to the evolution of physical theories, this comparison seems to us very appropriate, and it may be pursued in further detail.
>
> Whoever casts a brief glance at the waves striking a beach does not see the tide mount; he sees a wave rise, run, uncurl itself, and cover a narrow strip of sand, then withdraw by leaving dry the terrain which it had seemed to conquer, a new wave follows, sometimes going a little farther than the preceding one, but also sometimes not even reaching the sea shell made wet by the former wave. But under this superficial to-and-fro motion, another movement is produced, deeper, slower, imperceptible to the casual observer; it is a progressive movement continuing steadily in the same direction and by virtue of it the sea constantly rises. The going and coming of the waves is the faithful image of those attempts at explanation which arise only to be crumbled, which advance only to retreat; underneath there continues the slow and constant progress whose flow steadily conquers new lands, and guarantees to physical doctrines the continuity of a tradition.[7]

Duhem's image captures the fact that, while there may be a progressive trend, it is never inexorable. There are wrong turns, backtracks, and lulls, that often seem more impressive than the slow ground swell of change.

The second is the image of the building, constructed by many workers, each with their own task. This was the image presented by Vannevar Bush, a leader in the post-war development of science in the USA and creator of the National

Science Foundation. It is particularly interesting because it appears under the title of 'Endless Horizons'. This was Bush's rallying cry for the progress of science. It is a phrase that is often used in the USA when the question of the open-endedness of science is debated.[8] Bush begins by highlighting the mixture of order and disorder that characterizes scientific activity, with disorder especially evident to the outsider, and the way that the activity is self-organizing, rather like that of an ant colony,

> The process by which the boundaries of knowledge are advanced, and the structure of organised science is built, is a complex process indeed. It corresponds fairly well with the exploitation of a difficult quarry for its building materials and the fitting of these into an edifice; but there are very significant differences. First, the material itself is exceedingly varied, hidden and overlaid with relatively worthless rubble . . . Second, the whole effort is highly unorganised. There are no direct orders from architect or quarrymaster. Individuals and small bands proceed about their business unimpeded and uncontrolled, digging where they will, working over their material, and tucking it into place in the edifice.

He goes on to broach the question of whether science is invented or discovered: the building often uncovers pieces which seem so well adapted to match other quite separate parts that it seems as if each was fashioned to fit with the others,

> Finally, the edifice itself has a remarkable property, for its form is predestined by the laws of logic and the nature of human reasoning. It is almost as though it had once existed, and its building blocks had then been scattered, hidden, and buried, each with its unique form retained so that it would fit only in its own peculiar position, and with the concomitant limitation that the blocks cannot be found or recognized until the building of the structure has progressed to the point where their position and form reveal themselves to the discerning eye of the talented worker in the quarry. Parts of the edifice are being used while construction proceeds, by reason of the applications of science, but other parts are merely admired for their beauty and symmetry, and their possible utility is not in question.

He notices the curious sociology of the community of builders, workers, organizers, drones, and spectators:

> In these circumstances it is not at all strange that the workers sometimes proceed in erratic ways. There are those who are quite content, given a few tools, to dig away unearthing odd blocks, piling them up in the view of fellow workers, and apparently not caring whether they fit anywhere or not. Unfortunately there are also those who watch carefully until some industrious group digs out a particular ornamental block; whereupon they fit it in place with much gusto, and bow to the crowd. Some groups do not dig at all, but spend all their time arguing as to the exact arrangement of a cornice or an abutment. Some spend all their days trying to pull down a block or two that a rival has put in place. Some, indeed, neither dig nor argue, but go along with the crowd, scratch here and there, and enjoy the scenery. Some sit by and give advice, and some just sit.

He singles out a particular class of master builders with uncanny vision, who foresee what will work best, somehow sensing the structure that no one else can see, even though it is right in front of their eyes:

> On the other hand there are those men of rare vision who can grasp well in advance just the block that is needed for rapid advance on a section of the edifice to be possible, who can tell by some subtle sense where it will be found, and who have an uncanny skill in cleaning away dross and bringing it surely into the light. These are the master workmen. For each of them there can well be many of lesser stature who chip and delve, industriously, but with little grasp of what it is all about, and who nevertheless make the great steps possible.

And, finally, there are those who would seek to explain the building, its history, its meaning, and its beauty: all play a part in bringing the project to fruition.

> There are those who can give the structure meaning, who can trace its evolution from early times, and describe the glories that are to be, in ways that inspire those who work and those who enjoy. They bring the inspiration that not all is mere building of monotonous walls, and that there is architecture even though the architect is not seen to guide and order . . .
>
> There are also the old men, whose days of vigorous building are done, whose eyes are too dim to see the details of the arch or the needed form of its keystone, but who have built a wall here and there, and lived long in the edifice; who have learned to love it and who have even grasped a suggestion of its ultimate meaning; and who sit in the shade and encourage the young men.[9]

The third image that recurs in the attempts to understand the growth of science and mathematics is that of the tree. Unlike the tide or the building, it is a living thing, sprouting branches of different strengths, drawing strength through its roots to many sources. Karl Popper writes that

> we should have to represent the tree of knowledge as springing from countless roots which grow up into the air rather than down, and which ultimately, high up, tend to unite into one common stem.[10]

The nineteenth-century mathematician James Joseph Sylvester saw mathematics, including those parts which underwrote the scientific enterprise, as a vast growing tree of knowledge that could never come to an end, for

> mathematics is not a book confined within a cover and bound between brazen clasps, whose contents it needs only patience to ransack: it is not a mine, whose treasures may take long to reduce into possession, but which fill only a limited number of veins and lodes; it is not a soil, whose fertility can be exhausted by the yield of successive harvests; it is not a continent or an ocean, whose area can be mapped out and its contour defined: it is as limitless as that space which it finds too narrow for its aspirations; its possibilities are as infinite as the worlds which are forever crowding in and multiplying upon the astronomer's gaze; it is as incapable

of being restricted within assigned boundaries or being reduced to definitions of permanent validity, as the consciousness of life, which seems to slumber in each monad, in every atom of matter, in each leaf and bud cell, and is forever ready to burst forth into new forms of vegetable and animal existence.[11]

Some commentators are firm in their belief that the analogy with a living thing is quite distinct from that of the building. Here is Sir Michael Foster reporting to the Smithsonian Institution in 1899 on the growth of American science during the nineteenth century:

> The path [scientific progress] may not always be in a straight line; there may be swerving to this side and to that; ideas may seem to return again and again to the same point of the intellectual compass; but it will always be found that they have reached a higher level . . . Moreover, science is not fashioned as is a house, by putting brick to brick, that which is once put remaining as it was put to the end. The growth of science is that of a living being. As in the embryo, phases follows phase, and each member or body puts on in succession different appearances, though all the while same member, so a scientific conception of one age seems to differ from that of a following age.[12]

Yet, while the analogies might sound superficially different, at their heart they are the same. For they both fix upon an aspect of organized complexity that characterizes the building process in the same way that it characterizes the living process: many components working together to produce a totality bigger than the sum of their parts. The outcome is not something that will be understood by listing its ingredients or by isolating the activity of a single connection or building worker. It is what it is because of the intricate network of inter-relationships between its parts as much as through the identity of those parts.

There is another image of scientific progress which we can introduce. It is a new one, but captures some of the unpredictabilities of progress, and the interlinking of different developments in separate areas of science, in a new way. It is a model based upon the way a disease or a rumour might spread through a population. At any time, we could view our scientific knowledge as a collection of islands of information which are internally connected by measurements, theoretical connections, analogies, and so forth. The more of these cross-connections there are, the more tightly are these facts bound together by the requirements of self-consistency. This does not guarantee that they are all correct, of course, but it makes it harder to make an accretion of false information. Much of the everyday business of science involves the gradual expansion of these little islands of knowledge, deepening the interconnections between the ideas and facts within their encompass. Often, that progress is made, not by new discoveries, but by finding new ways in which to derive known things. These new derivations may be simpler in some way; not just by being briefer, but perhaps by using simpler combinations of ideas. (Ironically, this usually means they are longer!) Outsiders would be surprised by how much of the literature of science is composed of new ways of deriving things that we already know, or follow-up observations of some phenomena already observed by someone else. These confirmations strengthen the network of inter-connections within each island of knowledge, adding more strands to the bindings between different facts.

As the little islands expand in size, something more spectacular can occasionally occur. An insight or an observation might be made which allows one island to make contact with another. Powerful ideas allow interconnections to be made between many islands and, when they do, the body of connected ideas suddenly expands dramatically. This phenomenon, which scientists call 'percolation',[13] differs from diffusion of ideas (see Fig. 3.1). Its characteristic feature is a sudden jump in the size of the total region that is connected as the chance of making connections between individual facts slowly increases

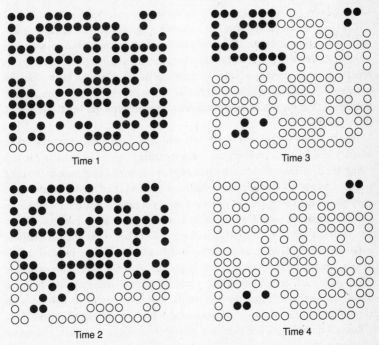

Fig. 3.1 The process of percolation. There is a probability that the white regions will infect their immediate neighbours. The characteristic of a percolation problem is the existence of a critical value of that probability which, when it is exceeded, ensures that the entire volume becomes infected.

towards a 'critical' level. This is what makes the spread of knowledge resemble an epidemic. If an orchard is blighted with a disease, the blight may begin with just one infected apple on a single tree. Gradually, other nearby apples on the same tree will become infected. Then, the blight can spread by jumping from tree to tree. The closer the trees are together the faster the blight will spread. If similar orchards are compared in which the trees are spaced differently, we can examine the probability that the infection spreads through the entire orchard in a given period of time. As the spacing is reduced, there is a point where this probability suddenly leaps up and complete spread is certain. Rumours similarly spread rapidly when enough people talk to one another frequently enough.

In practice, our knowledge is far from percolating into a completely linked system of ideas. Parts of biology have no good links with biochemical studies of the origin of life, and these in turn are connected only tenuously to our astronomical knowledge of how planets form. Computer science is trying to produce a significant overlap with studies of the brain, but so far the overlap is small and surprisingly weak: so much so that some people deny that it exists at all.

The most striking thing about this quest for percolating connections is why we are so keen to find them. For reasons of aesthetics, or inherited monotheistic religious belief in the unity of the Universe, we tend to believe that all things are related at some deep level. Any description of the Universe that used two ultimate principles would be regarded as inferior to one based upon a single principle. Unity tends to be viewed with greater approval than diversity in discussions of laws or foundations.

This tendency is captured by the percolation picture. Sometimes collections of ideas remain separate from the main spreading web of knowledge for very long periods of time. This can be because they are wrong in some fundamental way and so cannot be coordinated with other things that we know. But there have been other examples, like Einstein's theory of general relativity, which seem to have been discovered ahead of their time. Einstein's remarkable theory remained an island unto itself for a long time before experimental methods could allow it to develop by contact with astronomy and experimental studies of gravitation. Only recently has it begun to percolate with our studies of elementary particles and the other forces of Nature. Ironically, the bridging development that effected the percolation has been described by one of its creators, Ed Witten of Princeton University, as fifty years ahead of *its* time.[14]

One of the subtleties of Nature is that it is constituted in such a way that considerable progress can be made in advancing knowledge within separate islands of knowledge without the need for global percolation to have occurred. Science can make considerable progress by being reductionistic. The fashionable holistic picture of everything being required for any understanding of parts of the whole is not born out by our experience of the scientific method. Those Eastern cultures that adopted a holistic philosophy made little progress in science while they held fast to it. This is not to say that a holistic view does not have an important place in the understanding of things. It does: but only after some progress has been made understanding Nature piece by piece.

Another feature of this percolation picture is the expectation of sudden changes in the level of interconnectedness of our knowledge. A number of small advances—if they are in the right directions—can bring about a huge increase in our core of interrelated knowledge. Whereas small increments of steady progress are the hallmarks within the islands of specialized knowledge, leaps and bounds characterize the percolation of those islands with others. This is a different picture from Thomas Kuhn's idea of shifting paradigms accompanying radical change.[15] That picture fails to appreciate that there are reasons for the adoption of new theories and pictures that are not simply those which might characterize any change of fashion in collective activities, whether they be the Paris fashions, organized crime, or even hairstyles.

One of the things that this percolating view of scientific progress resonates with is our feeling for what constitutes a great advance. Great ideas unify superficially unrelated concepts. And it is here that our sense of beauty encompasses both the arts and the sciences. Beauty is the presence of unity in the face of superficial diversity. That unity can be in a pattern of ideas as much as in the pattern of petals on a flower or in the drawing together of the traits of character that define a tragic hero. There is a danger here as well. Crank science is full of vain attempts to find 'magic formulae' that derive the constants of Nature from other numbers, whether they be the dimensions of the Great Pyramid, the notes of the musical scale, or the decimal expansion of π. We have an instinct for unification that is part of what defines our intelligence. We can synthesize different facts, sort them into collections, see common factors, and thereby reduce the amount of information that is required to store and recall the information. This all goes to show that there is no magic formula, or definition, that defines good science.

Higgledy-piggledyology

Nothing you can't spell will ever work

WILL ROGERS

When considering where science might find itself heading in the far future, it is important to recognize a twofold thrust of scientific inquiry today. The quest of a subject like fundamental physics is to identify the most elementary building blocks of Nature and the laws that govern them. At present, it is believed that there are just four of these 'forces of Nature' and it is believed that these are not separate forces, but just different manifestations of a single 'super' force. The four forces—the strong, weak, electromagnetic, and gravitational forces—govern every physical phenomenon that has been observed in Nature. The mathematical theories that govern these forces are each of a special variety, known as 'gauge' theories, which take their structure from the requirement that a certain abstract pattern, created by the properties of the particles they govern, must be preserved by the law of Nature governing the action of the force in question. The quest for a grand unification of these different forces into a single theory is a search for a single overarching symmetrical pattern into which these four patterns can be embedded and united into a single picture rather like the pieces of a jigsaw puzzle. Some possible outcomes of this search will be discussed in Chapter 5. Here, our purpose is to highlight one vitally important point. Even if this small collection of laws of Nature is all that there is, and their unification is satisfactorily effected, there remains much to be done.

It is one thing to know the laws of Nature, but quite another to know the outcomes of those laws. The outcomes of laws of Nature are far more complicated than the laws themselves because the outcomes do not have to possess the symmetries of the laws. I am located at a particular place in the Universe at this moment, but the laws of Nature do not have any preference for particular places and times. They are entirely democratic. Rather, in any outcome of the laws those symmetries are broken or hidden. This simple fact is what allows our Universe to be governed, as it appears to be, by a very small number of simple symmetrical laws, yet display a vast array of complex asymmetrical states and structures. It also reveals why science is so difficult. We see the world of broken symmetries in the events and structures around us and have to work backwards to reconstruct the symmetrical laws which govern them.

This division of the scientific perspective into laws and outcomes helps us appreciate why some of the disciplines of science are so different in outlook. Ask the elementary particle physicists what the world is like and they may well tell you that it is very simple—if only you look at it in the 'right' way. Everything is governed by a small number of fundamental forces. But ask the same question of biologists or condensed-state physicists and they will tell you that the world is very complicated, asymmetrical, and haphazard. The particle physicist studies the fundamental forces with their symmetry and simplicity; by contrast, the biologist is looking at the complicated world of the asymmetrical outcomes of the laws of Nature, where broken symmetries and intricate combinations of simple ingredients are the rule. The observed structures are prevalent because they are the most persistent rather than the most symmetrical of possibilities.

If we refocus upon the future course of science we can imagine that our Universe might be quite simple at the level of the number of fundamental forces and the multiplicity of different elementary particles. We *might* be able to arrive at a logically consistent description of these forces. Sometimes this type of completion is referred to as a 'Theory of Everything'. But it is important to appreciate that this is something of a misuse of English. To the outsider 'everything' means what it says—everything, with nothing left out! But this is not what physicists mean. A Theory of Everything is intended to unite the different forces of Nature (at present believed to be just four in number). As a by-product, it ought to do some other impressive things at the same time. For instance, it could predict the identities of all the most elementary particles of matter, and even their properties as well. If it did the latter it should offer fairly clear predictions which would lay it open to observational test. But one must beware of expecting too much of such a theory. It is not an oracle which will print out an explanation of every single thing we see in the Universe, together with a list of all the other things we could see if we looked in the right places. There may be no limit to the number of different complex structures that can be generated by combinations of matter and energy. Many of the most complicated examples we know of—brains, living things, computers, nervous systems—have structures which are not illuminated by the possession of a Theory of Everything. They are, of course, permitted to exist by such a theory. But they are able to display the complex behaviours they do because of the ways in which their subcomponents are organized. It is one thing to have the Theory of Everything: quite another to find all (or even some) of its solutions. Nor is this an idle worry. The presently preferred candidate for a theory of this all-encompassing variety—string theory—appears to contain all sorts of information about the elementary particles of matter but, so far, no one knows how to solve the theory to extract that information. The mathematics is at present beyond us.

Thus, when we speculate about the fate of science we must be sensitive to the twofold nature of scientific progress. We can at least imagine that fundamental science might achieve its goal. (Later we shall look at some of the ways in which it might not.) It is not, however, so easy to imagine how the catalogue of the outcomes of those laws might be completed. It is this world of outcomes that fuels the growth of technology and applied science.

In recent years, developments in elementary particle physics have focused attention on the search for the ultimate 'laws of Nature'. This has led to several claims that the 'end of physics' might be in sight.[16] But no one has ever suggested that the end of the study of the outcomes of the laws of Nature might be in sight. In order to gain some overall perspective on the situation, it is useful to chart the state of different sciences in a diagram which plots the extent to which we have a good understanding of the underlying laws and equations that govern what goes on (the realm of the laws of Nature) versus the level of complexity witnessed in the outcomes of those laws (the world of the outcomes to the laws). As that complexity grows, so our understanding and ability to predict with accuracy what will happen in the future diminishes (see Fig. 3.2).[17]

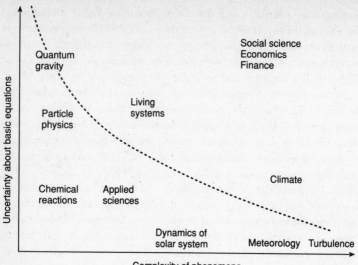

Fig. 3.2 A schematic representation of the degree of uncertainty that exists in the underlying mathematical equations describing various phenomena relative to the intrinsic complexity of the phenomena, after David Ruelle.

We can draw a dotted curve through the diagram shown in Fig. 3.2 which, in some sense, divides the subjects where we have a very good understanding of what is going on (and why) from those where we do not. Notice that it is possible to know the laws governing what you are studying, as in the study of highly turbulent liquids, and yet be in poor shape when it comes to explaining what is seen.

Selective and absolute limits

Hiding between all the ordinary numbers was an infinity of transcendental numbers whose presence you would never have guessed until you looked deeply into mathematics

CARL SAGAN[18]

When considering the limits that might exist to the future development of our knowledge about the physical world, in all its aspects, we need to distinguish some different breeds of limit. Suppose that all that could be known was laid out in a line of boxes stretching out in front of us. The line might be unending or it might have an end. Let us suppose it to be unending. Then there could be an 'absolute' limit upon our knowledge of the world in the sense that only a finite number of the boxes could be opened by us and by our descendants. Even though every unopened box might contain exactly the same information as the ones we have already opened, we can't know that until we opened them. So, strangely, we can never know that we know everything, even if we do.

Alternatively, we might encounter a further difficulty. The boxes might get smaller and smaller, and so more difficult to open, reflecting the greater effort required to extract the next piece of information about the world. At some point, we might encounter a box that was too difficult to open, perhaps for some deep reason to do with the character of the world itself, or a mundane one like the prohibitive economic costs. These would provide absolute and practical limits, respectively.

Another possibility is that we might have access to only one box in every ten of our never-ending line. Our exploration of Nature would then never be more than 10 per cent efficient, even though there would be no end to the number of things that we could discover. In this case, there are 'selective' limits to what we can know but not absolute ones. One can refine this picture even further. It would be possible for us to have an unending growth of knowledge which unveiled only an *infinitesimal* part of what could be known at that stage. If the unknowable things were densely packed like the collection of all decimal numbers (including unending ones) and what could be known was the collection of things labelled by the infinite list of whole numbers 1,2,3,4,5 . . . and so on, then we would always have missed finding an infinite number of things, even if we never missed any of the whole numbers in the list. This distinction between *selective limitations* on our ability to find out everything within some domain, like every variety of chemical molecule or every possible game of chess, and *boundaries* which we cannot cross, appears first in the writings of Immanuel Kant, who wrote that

> In mathematics and in natural philosophy, human reason admits of limits but not of boundaries, namely, it admits that something indeed lies without it, at which it can never arrive, but not that it will at any point find completion in its internal progress. The enlarging of our views in mathematics and the possibility of new discoveries are infinite: and the same is the case with the discovery of new properties of nature, of new powers and laws, . . . [19]

The most intriguing thing about the existence of selective limits is that we can be blissfully unaware of their existence. Absolute limits become apparent when no new fundamental discoveries are made for long periods.[20] By contrast, from a human point of view, scientific progress could appear to be accelerating (the 10 per cent of boxes that we open might always contain important new information), even though we were acquiring a smaller and smaller fraction of accessible information (the unopened boxes might contain even more!).

On reflection, this is a fair picture of the actual state of affairs in the past and present. Looking backwards, we can see how progress was invariably being made despite missing a huge number of things that we now know were accessible to investigators of the time, if only they had known where to look. At any moment of history, there are not merely questions one can ask but cannot answer, there are questions which there is no reason to ask. Whatever economic and human resources were made available to Pythagoras for the purposes of investigating the natural world, the results would have been rather shallow even by our own standards. He would not have known what questions to ask, nor could he have known. There is no reason to doubt that the present state of affairs is any different.

There is one important aspect of a future that is selectively limited that is of great practical importance. Although it may be possible to keep on learning fundamentally new things for ever, what is the *rate* at which we can learn them and what is the cost?

Will we be builders or surgeons?

Historians of ideas soon learn—to their dismay—that their subject appears to be mathematically dense: between any two people who wrote on the matter there appears to be another.

GRAHAM PRIEST[21]

There are two important roads to knowledge about the world. There is the path along which we progress by dissecting complicating things, breaking them down, step by step, into simple manageable pieces. This approach to Nature is sometimes called 'reductionism'. It allows explanations of complicated things to be 'reduced' to statements about what they are made of. This is also sometimes called the 'bottom-up' approach to Nature. Taken to extremes it would see human psychology reduced to biochemistry, biochemistry to molecular structure, molecular structure to atomic physics, atomic physics to nuclear physics, nuclear physics to elementary particle physics, and elementary particle physics to quantum fields or superstrings, and superstrings to . . . well, mathematics, perhaps? This approach plays an important role in our investigation of the world and has an obvious progressive aspect. The frontiers of elementary particle physics define the smallest scale at which we have been able to prise matter apart to discover what it is made of. But this is not the only route to understanding. Although it is extremely effective at arriving at an understanding of relatively simple things, it is less than helpful when applied to the most complex structures that we find in the world. Treacle is sticky and it will be found to be made of atoms, but we should not expect each atom to possess a little smidgen of stickiness.

Very complex structures have a general feature: they display complexity because of the intricate organization of a very large number of simple components. Whether that structure be an economy, a weather system, a liquid, or a brain, it is what it is and does what it does because of the way in which its constituent parts are organized, not primarily because of what they are. All the examples we have listed are made of atoms if you look at a low enough level, but that does not help us to understand the distinction between a book and a brain.

Complex structures seem to display thresholds of complexity which, when crossed, give rise to sudden jumps in the complexity. Take groups of people. One person can do many things; add another person and a further relationship becomes possible; but gradually add a few more people and the number of complex interrelationships grows enormously. Economic systems, traffic systems, computer networks: all exhibit sudden jumps in their properties as the number of links between their constituent parts grows. Consciousness is the most spectacular property to emerge in this way when a very high level of complexity is reached in a connected logical network, like the brain.

The division between symmetrical laws and complex outcomes is often reflected in the way science organizes itself. Some subjects, like biology, are exclusively involved in the study of the messy world of complex outcomes, while others, like particle physics, largely focus upon the pristine symmetries of Nature's fundamental laws. The skills of the scientists involved in these different enterprises are quite different. Occasionally there is an attempt by one group to apply their expertise in another area. This is interesting. Attempts to understand consciousness provide an intriguing example of the different psychologies of two branches of science. The biologists and neurophysiologists are used to dealing with complicated natural structures which emerge in a messy way as the result of a historical process of accidents and natural selection. They expect that a complex phenomenon like consciousness will be explained as the outcome of a huge number of mundane processes organizing themselves over a long period of time into a structure which learns in the way that a neural network does: that is, consciousness is like a computer system that 'evolves' by a microscopic version of natural selection. A typical example of this type of messy explanation in which symmetry or simplicity plays no necessary role, where it is simply persistence and marginal advantage over alternatives that win out in the long run, is provided by Edelman's picture of Neural Darwinism.[22] Here, the development of the brain's networking is a constantly evolving entity in which useful, much used connections are reinforced at the expense of those that are less used.

Fundamental physicists reveal a quite different bias. In their subject the deepest and most important things are the basic mathematical structures behind the laws of Nature. Physicists expect 'important' things to be the most fundamental; and 'fundamental' means simple, symmetrical, or mathematically subtle. It doesn't mean the survival of the fittest. As a result, physicists find it very hard to imagine that anything they judge to be fundamental could have a haphazard messy explanation rather one that follows from some elegant requirement of mathematical symmetry. Of course, physicists think that consciousness is fundamentally important and worthy of explanation, and so some of them tend to think it can't have one of these complicated and messy explanations.[23] Instead, to the astonishment of the biologists,[24] they introduce things like quantum gravitation and intrinsic non-computability at the microscopic level in order to explain macroscopic features of the mind.

The futures market

Never say never again.

JAMES BOND

The easiest way to get us thinking about the possible future for science is to consider two aspects only: whether or not there is an unlimited store of fundamental information about Nature to be uncovered, and whether or not our capabilities are limited or not. This makes possible four distinct futures:

Type 1 future: Nature unlimited and human capability unlimited;
Type 2 future: Nature unlimited and human capability limited;
Type 3 future: Nature limited and human capability unlimited;
Type 4 future: Nature limited and human capability limited.

Before we explore these future possibilities in more detail it is important to bear some general points in mind. When we consider the option that there might be only a finite number of things to learn about Nature we are talking not about the number of different things that Nature manifests—there might be no limit to the number of galaxies in an infinite universe—but of the basic principles and 'laws' that seem to allow us to characterize whole collections of individual entities in Nature. Actually, this restriction to finiteness is not as sweeping as it first appears. We tend to think of the number of possible snowflakes, the number of possible musical works, or the number of genetically possible human beings, as being 'unlimited' in our casual use of this word. But in each of these cases the number of possibilities is not unlimited: it is a huge, but none the less a finite number.

If there are an infinite variety of distinctive forms of complexity, then we are faced with an insuperable challenge. The philosopher of science William Kneale worries about the prospects of capturing everything in the complex world of outcomes,

If by the 'infinite complexity of nature' is meant only the infinite multiplicity of the phenomena it contains, there is no bar to final success in theory making, since theories are not concerned with particulars as such. So too, if what is meant is only the infinite variety of natural phenomena . . . that too may be comprehended in a unitary theory . . . Nor does it help to say that there is indeed a true explanatory theory in some Platonic heaven but that it is infinitely complex and so not to be comprehended by men. For if there can be an infinitely complex proposition, it will certainly not be a single explanatory theory in any ordinary sense of that phrase, but at the best an infinite conjunction of explanatory theories. Perhaps we can produce successive approximations to such a conjunction, if there is nothing else to work for, but in that case our best hope of success will be by steady accumulation of the separate items than by perpetual revolution.[25]

When we talk about the future of human capabilities we need not restrict ourselves to that of unaided human investigation. Just as we are able to do 'super'-human tasks by using fast computers, so in the far future we can expect forms of artificial intelligence which will do much more than simply increase the speed of human calculation or the quantity of data that can be assembled and compared at one time. Ultimately, the bulk of the scientific enterprise might be pushed forward by forms of machine intelligence that are able to extend human capabilities in both predictable and unpredictable ways.

In mapping out the course of events which might lead to each of our four futures it is useful to introduce a graph of the change of knowledge with time (Fig. 3.3). The line charts the increase of knowledge about the Universe. The region above the curve of progress is the unknown; that below it is the known. The total amount of accumulated knowledge is the area underneath a curve of progress. In each case, we should bear in mind that human investigation may

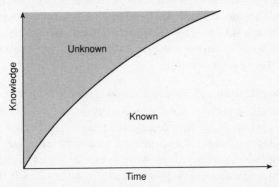

Fig. 3.3 Curve of progress. A schematic representation of the growth of knowledge with time. The curve separates the known from the unknown.

not be an activity that continues indefinitely (regardless of whether or not it is limited in its ability to discover things about the visible Universe). It is easy to envisage futures in which human life disappears.[26] Already, we can see that we have gone perilously close to having wars that escalate into full-scale nuclear exchanges; industrial pollution threatens the climatic stability of our planet; energy sources are steadily being used up and non-fossil fuels create new environmental risks; ice ages return every hundred thousand years; chance encounters with meteors, comets, and asteroids are a constant menace to planet-based life; disease or corruption of our staple food sources are a constant threat. Existence is precarious: as the world becomes an increasingly sophisticated technological system, it is increasingly at risk from the consequences of its own headlong rush for development.

Pondering these things, it is not difficult to imagine that it might be very difficult, or even impossible, for civilizations to persist for too long after they become industrialized. A critical level of technical knowledge may lead inevitably to the gradual (or sudden) extinction of its possessors. If so, any extremely long-lived civilization is likely to be very unusual and, I believe, therefore qualitatively (not just quantitatively) different from our own. We should remember that we risk being unrealistic if we extrapolate human (or even superhuman) progress indefinitely into the future. Even if civilizations do not self-destruct, they ultimately face environmental crises of cosmic proportions as stars run out of nuclear energy and galaxies disintegrate. They may even face an implosion of the entire universe into a Big Crunch in which conditions mirror those of the Big Bang.[27] We shall ignore these background problems except in as much as they provide long-lived civilizations with challenging scientific problems. Their survival depends ultimately upon their ability to come up with solutions. However, our own societies have lessons for us here. It is very difficult to get politicians and democracies to plan for the far future. There are enough problems for today and tomorrow, let alone those thousands of years to the future. What sort of society would come to invest a huge part of its intellectual and material resources to cope with problems tens of thousands of years in the future?

A Type 1 future: Nature unlimited and human capability unlimited

A Type 1 future seems to be a simple extrapolation of our past and present experience. New discoveries keep coming, and bring with them new problems as well as solutions to old ones. Such progress need not be inexorably upwards; there can be dark ages when progress slows, or even declines, and there can be great surges brought about by the insights of an Einstein or a Darwin (Fig. 3.4).

Wiggles in the curve of progress reflect characteristic intervals of time, like the length of human lifetimes, of particular schools of thought, and of the social

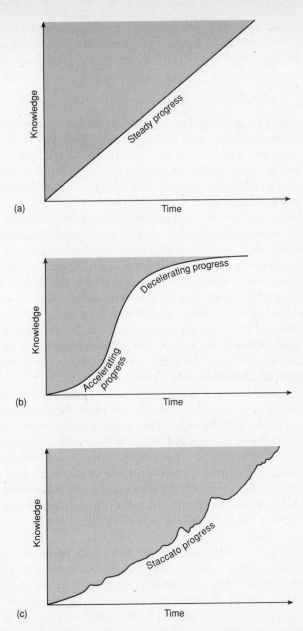

Fig. 3.4 Three patterns of indefinitely expanding knowledge: (a) steady expansion; (b) variable expansion: initial acceleration is replaced by deceleration; (c) an overall trend of expansion modulated by small fluctuations.

environments in which they work. However, just because progress is unlimited does not mean that the cost of acquiring information stays the same. A sequence of terms which get smaller and smaller can still add up to an infinite total: for example, consider the series

$$1 + 1/2 + 1/3 + 1/4 + 1/5 + 1/6 + 1/7 + 1/8...ad\ infinitum$$

Although each term is smaller than the previous one, the total sum never just gets closer and closer to any limit. Its sum can exceed any number you care to specify so long as you take enough terms.[28] We say this is a 'divergent' series. Thus, if each term in the series were to mark the progress made during each decade in the future, the total progress would never be bounded but the rate of progress would become ever slower.

This illustration shows that unlimited progress does not necessarily mean accelerating progress. The curve of progress might become shallower and shallower. Important new scientific discoveries might eventually occur much less frequently than one in a typical human lifetime. The onset of a situation like this might prove a serious disincentive to continued investigation. Other cerebral pursuits with a more rapid turnover of novelty, employment, and satisfaction might prove much more attractive.

The defence of this long-term future requires us to take seriously two extrapolations. We need to consider whether Nature is likely to offer an unlimited number of important things for us to discover and whether we should expect our capabilities to be unlimited.

Until quite recently most scientists would have felt uncomfortable with the idea that Nature might be an exhaustible store of riches. But developments in elementary particle physics have opened up a vista in which the underlying laws of Nature—the elementary particles of matter, the forces that govern them, and the interactions between them—might be few in number and so constrained by the requirements of logical consistency that they might exhibit only a few special forms. Elementary-particle physicists would not be surprised if a time-travelling tourist from the future were to tell them that there are only four basic forces of Nature and one particular superstring theory describes all their workings.

This particular aspect of the scientific endeavour might be one that could be completed. As we have looked more deeply into the structure of the underlying laws of Nature the impression we have is that things are very often simpler than we might have suspected. Just as the most expert computer programmer is the one who can write the shortest program to effect a particular task, so we might expect the Architect of the ultimate program that we call the laws of Nature to be elegantly economical on logic and raw materials. It is a common tendency to think that it would be a hallmark of the Universe's profundity if it were unfathomably complicated, but this is a strange prejudice. This view is motivated by the idea that the Creator needs to be superhuman—and what better way to assert that superiority than by incomprehensibility? But why should that be so? Anyone can explain how to assemble a model aircraft in 500 pages of instructions; it is not so easy to do it in 10 lines. Profound simplicity is far more impressive than profound complexity. The most remarkable thing about the Universe might ultimately turn out to be the very small number of rules and components required to define it. The alternative is that there is a bottomless complexity: each advance into the realm of the very small revealing a new world of structure, each significant gain in our ability to measure faint forces revealing new, previously hidden, effects. The late David Bohm was attracted to this bottomless well of information as the hallmark of Nature:

> Generally speaking, by finding the unity behind the diversity, one will get laws which contain more than the original facts . . . the whole scientific enterprise implies that no theory is final . . . at least as a working hypothesis science assumes the infinity of nature; and this assumption fits the facts much better than any other point of view that we know.[29]

Joining Bohm in this limitless vision of Nature is Eugene Wigner, one of the greatest physicists of the twentieth century. Wigner sees Nature composed of layers of complexity which, like onion skins, reveal layer after layer beneath. To penetrate those layers of reality, we shall need to develop deeper and deeper concepts. Without saying whether we are faced with an infinite sequence of levels or a finite one, Wigner sees no reason why we should be able to peel away all the conceptual barriers to ultimate understanding:

> in order to understand a growing body of phenomena, it will be necessary to introduce deeper and deeper concepts into physics and this development will not end by the discovery of the final and perfect concepts. I believe that this true: we have no right to expect that our intellect can formulate perfect concepts for the full understanding of inanimate nature's phenomena.[30]

When we turn from the search to understand the make-up of the Universe and the rules that appear to govern it, we find a different situation. There is no end to the devices that we can build by joining together atoms and molecules (and perhaps even sub-nuclear particles, like quarks) in complicated patterns. Charles Babbage, the nineteenth-century inventor of the calculating machine, saw the self-perpetuating possibility that technology creates:

> Science and technology are subject, in their extension and increase, to laws quite opposite to those which regulate the material world . . . [the] further we advance from the origin of knowledge, the larger it becomes, and the greater power it bestows upon its cultivators, to add new fields to its dominions [and] . . . the whole, already gained, bears a constantly diminishing ration to that which is

contained within the still more rapidly expanding horizon of knowledge . . . it may possibly be found that the dominion of mind over the material world advances with an ever-accelerating force.[31]

Need there be any limit to constructable complexity? Not as far as we know, although, as we shall see, there will surely be limits on our ability to construct devices and networks which exploit those complexities to the full. Such projects require time and resources. They will be pursued only if there are very good reasons to do so.

A Type 2 future: Nature unlimited and human capability limited

The Type 2 scenario (Fig. 3.5) requires less of a twist of our imagination to accommodate. It is the picture that would probably gather most support from a random poll of individuals. It is certainly the most modest position to take. It respects Nature's diversity while recognizing our own limitations (Fig. 3.5).

Although our ability to make new discoveries might be bounded, this does not mean that our knowledge cannot still continue to increase for ever. But it would approach closer and closer to a limit imposed by one of many possible restrictions: the nature of our brains, the lack of materials and energy, our size, for instance. In this scenario, the limit would never actually be attained no matter how long our investigations continued.

Alternatively, we might reach our limit in a finite time. Subsequently, the cost of going further might be impossible for us to meet or we may have achieved some fundamental limit on the process of observations, information storing, or processing speed.

In the past, the view that our knowledge of the world will halt sooner rather

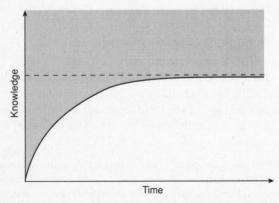

Fig. 3.5 A Type-2 future. Knowledge is always increasing, but is limited, while Nature is unlimited.

than later was surprisingly common. Denis Diderot, a prominent French intellectual of the eighteenth century, wrote in 1875:

> I dare virtually to guarantee that before one hundred years have passed one will not find three great mathematicians in Europe. That science will come to a dead stop pretty much where a Bernoulli, an Euler, a Maupertuis, a Clairaut, a Fontaine, and a D'Alembert and a La Grange have left it. They have erected the pillars of Hercules beyond which there is no voyaging.[32]

About a century later, on the other side of the English Channel, the scientist George Gore gave a fuller account of these possibilities, asking first whether Nature might not be finite in structure:

> Although we know but little of the actual limits of possible knowledge, there are signs that nature is not in every respect infinite. It is highly probable that the number of forms of energy and of elementary substances is limited . . . Not only does it appear highly improbable that an unlimited variety of collocations of different atoms, united to form different substances, can exist; but many combinations and arrangements of forces are incompatible, and cannot co-exist. From these considerations, therefore, there is probably a limit to . . . the amount of possible knowledge respecting them. The number of laws also which govern a finite number of substances or forces must themselves be finite.[33]

Gore then went on to voice his suspicion that human knowledge will for ever lag behind the challenge presented by Nature:

> The future limits of human knowledge seem to be infinitely distant . . . Our knowledge is finite, but our ignorance is nearly infinite . . . The amount of discovery in the future appears likely to be vastly greater than that of the past . . . [since] the whole realm of attainable knowledge appears immensely great in comparison with the powers of the human mind, the unfolding of it will probably require an almost infinite amount of labour, and therefore a vast period of time . . . how far man, with his finite intellect, will in future be able to explain the phenomena belonging to the various parts of the universe, and successfully predict effects, no one at present can even guess.

Unless, of course, there is some hidden anthropocentric design in Nature which matches its complexity to that of the sentient beings that exist within it:

> It is, however, reasonable to suppose that as the whole of nature is systematically framed in accordance with intelligent design, nothing in it is essentially inscrutable to intellectual powers, and the vast expanse of truth which remains unknown is only temporarily inscrutable, until the prior knowledge necessary to its discovery is obtained. And as ceaseless activity is a necessary condition of human existence, we may also conclude that new and improved intellectual processes of research will be invented, and that the entire universe of scientific truth will [ultimately] be investigated and discovered.

In Gore's closing lines, we see a hope that the questing human spirit will overcome all barriers and conquer the future, making intellectual conquests to match our evolutionary history of terrestrial exploration and discovery.

A Type 3 future: Nature limited and human capability unlimited

If the fundamental laws of Nature, and the principles which govern the organization of matter and energy into complex structures and configurations, are finite in extent, then unlimited capabilities would be sufficient to uncover them all. At some stage, in certain important respects, we would complete the scientific enterprise: all fundamental discoveries would have been made (Fig. 3.6). All that would remain would be more refined measurements. There would be new facts to gain but they would be mere details, 'further decimal places', upon which no fundamental theory would stand or fall; scientific papers might report that existing theories were confirmed to new levels of precision, but there would be no more surprises. Of course, we could never be sure, but as time went on enthusiasm would wane. Creative minds would look elsewhere for new challenges. Perhaps the design of other more complex virtual universes might turn out to be more interesting than the study of our own.

The late Richard Feynman was, perhaps reluctantly, attracted by this view. His work on elementary particle physics exposed him to a system of the world governed by a very small number of laws and basic forces. It would be interesting to know whether his view changed many years later after his work on complex computational systems. He wonders,

> What of the future of this adventure? What will happen ultimately? We are going along guessing the laws; how many laws are we going to have to guess? I do not

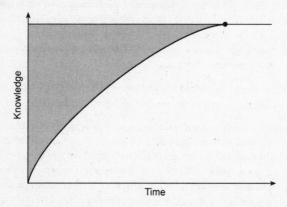

Fig. 3.6 A Type-3 future. Nature requires a finite amount of information for its full understanding and this lies within human capability.

know. Some of my colleagues say that this fundamental aspect of our science will go on; but I think there will certainly not be perpetual novelty, say for a thousand years. This thing cannot keep on going so that we are always going to discover more and more new laws . . . It is like the discovery of America—you only discover it once. The age in which we live is the age in which we are discovering the fundamental laws of nature, and that day will never come again. Of course in the future there will be other interests . . . but there will not be the same things that we are doing now . . . There will be a degeneration of ideas, just like the degeneration that great explorers feel is occurring when tourists begin moving in on a territory.[34]

Another American scientist, Bentley Glass, in his lecture about the issue of the whether science has the 'endless horizons' advertised by Vannevar Bush, stresses the distinction between fundamental discoveries, which may well be exhausted, and the filling in of the details—a secondary scientific activity—which might well go on for ever:

What remains to be learned may indeed dwarf the imagination. Nevertheless, the universe is closed and finite . . . The uniformity of nature and the general applicability of natural laws set limits to knowledge . . . We are like the explorers of a great continent who have penetrated to its margins in most points of the compass and have mapped the major mountain chains and rivers. There are still innumerable details to fill in, but the endless horizons no longer exist.[35]

A Type 4 future: Nature limited and human capability limited

A Type 4 future is the most complicated of eventualities. There are three principal possibilities, as shown in Fig. 3.7.

By an absurd coincidence the two limits shown in the text-figure might coincide so that we were just able to learn everything knowable in either a finite or an infinite time (Fig. 3.7(a)). This would imply some cosmic conspiracy, with ourselves the object of it, of the sort about which Gore speculated. More realistically, we would expect the two finite limits to be quite different. The limit on our capabilities might be at such a high level that it allowed us to determine all the fundamental principles governing Nature (Fig. 3.7(b)). This would be a very unusual self-referential situation. It would mean that the brain (or its artificial intellectual successors) would need to have greater complexity than the entire collection of principles governing all possible forms of organized complexity. This might well be impossible. More likely is a situation in which our capabilities fall significantly short of the capacity of Nature (Fig. 3.7(c)). This is a situation that reduces, essentially, to that of the Type 2 future introduced above.

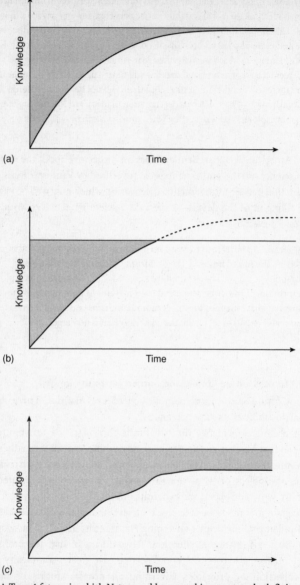

Fig. 3.7 A Type-4 future, in which Nature and human achievement are both finite, and can have different forms: (a) by coincidence, both limits are the same, and human investigation can reach Nature's limit after a finite (or perhaps only an infinite) time; (b) Nature's capacity is attained after a finite time because humanity's limit (dotted) exceeds it; this is like a Type-3 future; (c) human achievement falls short of Nature's capacity; this is like a Type-2 future.

How many discoveries are there still to be made?

Discovery consists of seeing what everybody has seen and thinking what nobody has thought.

ALBERT VON SZENT-GYÖRGYI[36]

Strange as it may sound, we can make some quantitative statements about the number of fundamental discoveries that science can be expected still to make. The problem is similar to that of proof-reading an article for typographical errors. Suppose that two editors, Jack and Jill, independently read a long newspaper article supplied by one of their journalists. Jack finds A typing errors, whilst Jill finds B typing errors. They compare copies and discover that they both found the same error on C occasions. How many errors do you expect to remain, unfound, in the article?[37]

Let us suppose that the total number of errors in the article is E. This means that the number that have yet to be found equals $E - A - B + C$. The last factor of $+C$ is so that we don't double count the errors that Jack and Jill both found. Now, if the probability that Jack spots an error is p, and the probability that Jill spots an error is q, then we expect that $A = pE$, $B = qE$, and $C = pqE$, because they search independently. So, $AB = pqE \times E$; hence $AB = CE$. Now we have the answer: the number of unfound errors equals $E - A - B + C = AB/C - A - B + C$, where we have replaced the unknown quantity, E, by AB/C. rearranging our formula, we have shown that the number of unfound errors is equal to $(A - C)(B - C)/C$; that is,

Number of unfound errors =

$$= \frac{(\text{Number found } only \text{ by Jack}) \times (\text{Number found } only \text{ by Jill})}{(\text{Number found by both Jack } and \text{ Jill})}$$

This result makes good sense. If Jack and Jill both found lots of mistakes, but neither of them found the same mistakes, then they are not very good proof-readers and there are likely to be lots of other mistakes that neither of them found.

What has proof-reading got to do with the future of science? It is clear that the same type of reasoning could be applied to the question, 'how many scientific discoveries are still to be made?' Instead of independent proof-readers, we would consider separate ways of investigating Nature; for example, astronomical observations in different wavebands, or particle physics experiments in different energy ranges. We would then ask how many fundamental discoveries have been found by separate investigations alone, and how many by more than one. The formula that we have given can easily be generalized to any number of independent investigations, and will give an estimate of the number of fundamental discoveries to be made without our needing to know the values of p and q, that is, how likely any type of investigation is to make a discovery.[38] Whether or not one wants to attempt to put numbers in these formulae, their value is that they reveal how the state of science should be judged by the extent to which its discoveries are reinforced by different lines of investigation. Since Nature is a deeply entwined unity, the extent to which we are able to repeat discoveries by completely different observational techniques gives us some insight into the depth at which we are probing the structure of the Universe. During the past fifteen years, the development of the links between particle physics and astronomy provides a remarkable example of this interrelationship. Many of our expectations about the structure of the Universe, like the fact that it contains a large quantity of non-luminous matter, are the results of independent lines of inquiry from both particle physics and cosmology. When this happens we are justified in believing that there are fewer new discoveries to make than had we found ourselves investigating quite separate types of predictions from the two lines of inquiry.

Summary

The will is infinite and the execution confined . . .
the desire is boundless and the act a slave to limit
WILLIAM SHAKESPEARE[39]

There are many images of science and the activities of scientists. Some imply that science will end, while others create an expectation of endless horizons. We have seen a variety of images of the progress of science: as a growing tide, a construction project, a living thing, or a percolating epidemic of interconnected knowledge.

We have come to appreciate a dual aspect of the Universe: the laws of Nature and the outcomes of those laws. The laws are few and simple, but the outcomes are numerous and complicated. When we consider the meaning of a so-called 'Theory of Everything', that physicists seek, we have to distinguish very carefully between finding the laws of Nature (this is the 'Theory of Everything'[30]) and understanding the complex outcomes of those laws. With this distinction established, we looked at some different types of limits to scientific progress, and went on to sketch some simple alternative futures. The relation between the information contained in Nature and the information that we are able to discover by observation and reasoning determines four coarse-grained futures.

CHAPTER 4

Being human

Human beings know a lot of things, some of which are true, and apply them.
When we like the results, we call it wisdom.

HERBERT SIMON

What are minds for?

The most important thing to realize about systems of animal communication is that
they are not expected to be systems for the dissemination of truth. Instead, they are
expected to be systems by which individual organisms attempt to maximize their
fitness by communicating to others things that may be true or false.

ROBERT TRIVERS[1]

Are there limits on our ability to understand the Universe which are imposed by the nature of our minds? This sounds like a serious possibility. Human minds have histories—long and tortuous histories. Like all other human organs they have made their way from the past to the present by an erratic path of trial and error. Small random variations have been sifted by their ability to aid survival and fecundity. All our present abilities are inheritances from the past. If, as we believe,[2] they cannot be pre-programmed for purposes in the far distant future, then they are likely to be far from optimal when used in our future quest to understand the Universe.

Many of our human attributes have obvious survival value. Language is extraordinarily advantageous.[3] But others are not so obviously helpful. Why do we yawn? Why do we have ear lobes? Why do we like music? As we ponder such questions, we must recognize that in some cases we have inherited attributes that were useful long ago in the ancient environments in which our earliest ancestors lived for very long periods of time. However, we also possess attributes which are just by-products of others. This means that many of our impressive mental abilities might not be the direct results of natural selection acting to promote the inheritance of that specific ability. They might be merely by-products of other adaptations to environments that no longer exist.

The human brain is the most complicated thing that we have encountered in the Universe. It weighs about three kilograms, not much more than a large tin of motor oil, but within that small mass lies a staggeringly intricate network of interconnections between a hundred billion neurons. It takes in information about the body, about the environment, controls limbs, and stores away information in ways that still remain a mystery. It learns; it remembers; it forgets; it dreams; it creates. Fortunately, the mystery is not complete. Brains have things in common with manmade computers. They possess an ability to 'run' different programs of all sorts ('software') that don't come built in. We can learn to play chess, do long division, or carry out all sorts of other very specialized activities. But underlying this flexibility there is a built-in system, akin to the hard-wired ROM of our home computer, that imparts the ability to run these other programs and defines our overall capabilities, speed of thought, and learning abilities.

The performance of the human mind is so impressive that we are apt to be misled about its imperatives. The largest supercomputers that we have constructed so far pale into insignificance compared with the complexity, flexibility, and compactness of the human brain. Supercomputers can always outperform the brain in specific abilities—notably speed in performing simple repetitive tasks—but pay the price in their lack of adaptability and their inability to learn about themselves. A good example of specialized computer skill was the chess matches between Deep Blue, an IBM chess-playing computer capable of examining 200 million positions per second, and the world champion Gary Kasparov, who is probably the strongest player ever. In the first contest (in 1996), after drawing the first game unexpectedly, Kasparov recovered the situation by adopting a style of play that required a level of overall pattern appreciation (which we would call strategic 'intuition') that Deep Blue was clearly lacking. Eventually, Kasparov won easily by 3 games to 1 with 2 drawn. In 1997, the new edition of Deep Blue was far stronger. Kasparov played badly, and was surprisingly beaten. In the future, Deep Blue is likely to be stronger still.

Deep Blue is significantly superior to all previous chess-playing machines and would make mincemeat of most human players in this rather specialized activity. An interesting example of a simple chess problems which defeat chess programs is shown in Fig. 4.1. The growth of computers' capability is shown in Fig. 4.2.

The amazing all-round performance of the brain in conducting feats of logical reasoning, doing mathematical calculations, and understanding esoterica from quarks to quasars can easily seduce us into thinking that this is what the brain is 'for'.

In my study I have two large volumes which contain reproductions of all Salvador Dali's paintings and drawings.[5] The volumes have been beautifully produced in a boxed set by the publishers, and contain a wealth of fascinating historical information about the artist, his life, and works. The aim of the editors was to produce a work that would display Dali's work to scholars and to

Fig. 4.1 Chess configuration with white to play and force a draw. While simple for humans, this was not so easy for the 1993 chess-playing program *Deep Thought*. Black is numerically in a far stronger position than white, but white can avoid defeat by moving the white king back and forth behind the impregnable line of white pawns running across the board. This situation is obvious to any human player, but *Deep Thought*, playing white, immediately made the mistake of capturing the black rook with a white pawn. This breaches the line of white pawns and creates a hopeless situation for white.

others with a more casual interest in his work. Yet, I find these heavy, boxed volumes make wonderfully robust book-ends—better than any purpose-made book-ends that I could find. This is not an entirely uncommon state of affairs: we discover that something designed for one purpose happens to be useful for others that were entirely unforeseen, even unforeseeable, when it was made. These unplanned uses are by-products of the one that its designers originally had in mind.

Not surprisingly, living things are often beneficiaries of dual uses as well. The hand was not evolved in order to embroider a tapestry, or to engineer Swiss watches; the ear was not primarily endowed with the ability to identify musical pitch. Yet, the hand and the ear can be remarkably adept at both.

We have learnt that our physical abilities are the result of a long process of adaptation to local circumstances in the presence of rival competitors. The bundle of attributes that has the greater chance of surviving in the long run will prevail. The 'long run' here is the entire history of humans and their pre-decessors, and extends over many millions of years. Our recent history, although marked by remarkable rates of progress, is a mere drop in this ocean of time. Despite our preoccupation with the rational products of the human mind—with science, technology, mathematics, and computers—such things are new and novel activities for us. Our aptitudes for them must be considered

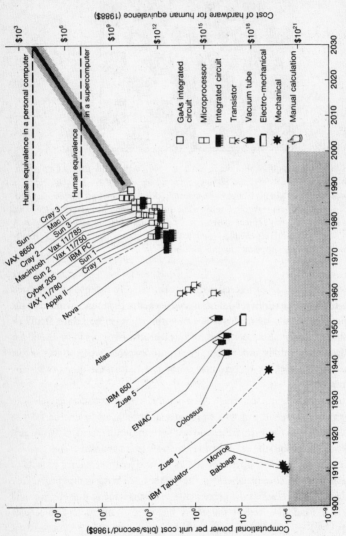

Fig. 4.2 Relative performance of the brain versus large computers. This graph charts the growth of computational power with time during the twentieth century. Human equivalence is shown, together with the declining cost of the hardware required for human equivalence in US dollars at 1988 prices. (Based on data gathered by Hans Moravec, *Mind Children*, Harvard University Press (1998).)

as by-products of other more basic abilities which the mind possesses because they produced a better chance of surviving in the distant past than the alternatives available at that time, when the environment may have been quite different in some respects from that of today.[6] The rate at which evolutionary change occurs in humans is far too slow to have made much difference to our innate abilities over the whole period which encompasses our recorded history (about 9000 years or so).

This simple fact has huge consequences. If our minds evolved primarily to deal with a sequence of complicated environments that our distant ancestors faced for millions of years, then that process will have endowed our minds with particular biases that were appropriate for dealing with the problems they faced. Those problems did not include passing examination papers in particle physics, or elucidating the mathematics of symmetry. But such esoteric things might well arise as by-products of something more basic which endows selective advantages. In Chapter 1 we introduced the survival value imparted by an appreciation for symmetry in our ancestors millions of years ago. Inanimate things tend to be asymmetrical but living things generally display left–right symmetry. They don't usually have up–down symmetry because of gravity, and do not have front–back symmetry if they are able to move. By detecting symmetry in a confused scene, one picks out potential predators, mates, and meals looking straight at you. Such an awareness provides a significant marginal advantage over those who do not possess it. The genetic cocktail that gives rise to it is therefore likely to survive, while that which leads to its absence will fall by the wayside because this means you will be at greater risk from predators, less likely to find mates, and more likely to go hungry. Moreover, this way of looking at things implies that an oversensitivity to patterns might well be a better thing to have than an undersensitivity. Better to be thought a little paranoid now and again for crying wolf than to miss the big bad wolf when he's really hiding in the bushes.

We inherit the consequences of an adaptation like this. Our sensitivity for pattern in overcrowded, or nearly featureless, scenes is aroused in all sorts of modern situations. We look at Mars through a telescope and see 'canals' on the surface. There is even a group of Americans who claim that there is a human face drawn out on the Moon's surface.[7] When our ancestors looked at the stars they saw all sorts of patterns: from ploughs and serpents, to bears and hunters. Modern astronomers are no less susceptible to this tendency: the Egg Nebula, the Crab Nebula, the Horsehead Nebula, like many others, are all named for their suggestive visual appearance. In all these cases, we see how our mental faculties are influenced by their past history. We inherit biases towards seeing certain things better than others.

It is important to appreciate that the abilities our minds and bodies possess were originally solutions to problems posed by environments in which we no longer live. Some adaptations to those past environments are still with us, but many are not. But it is also important to realize that they need not be optimal. Many people, even scientists who should know better, have been so seduced by the amazing intricacy of the adaptations that living things possess, that they assume they are perfect adaptations. But this is far from the truth. The human eye is a remarkable optical instrument, but it is far from the best possible.[8] Honey bees make very effective use of their raw materials in making a honeycomb but mathematicians know that they could be more efficient still.[9] This is hardly surprising. Perfect adaptations to environmental conditions can be prohibitively expensive. Any surfeit of resources invested in them must be paid for by more imperfect adaptation in some other department. What is the point of buying a very expensive set of 100-year guarantee spark-plugs for your car? None at all, when the rest of the car will have failed long before the spark-plugs.

This means that there is no reason to expect the human brain to be the best possible all-purpose reasoning instrument. It needed only to do better than similar alternatives into which it could evolve by small changes, and to outperform the brains of rivals. This it managed to do despite its evident fallibilities of memory and reasoning ability and the curious fact that it makes use of just part of its total resources.

This realization is an important consideration when we consider the future potential of human scientific investigation and whether limits imposed by our humanity restrict our knowledge of the Universe in significant ways. If our minds are fallible instruments we need to inquire more carefully into how their intrinsic limitations might bias and restrict what we can know of the physical world around us.

Scientists rarely consider the limitations and biases of the mind to be important. They see the human mind as a collection of problem-solving abilities which can be applied to any complex problem. Aided by fast computers, they believe that its logic will prevail given enough time. Some philosophers take a radically different view. They see the evolutionary heritage of the human mind as a guarantee of its fallibility. The most notable supporter of this postmodernist view is the American philosopher Richard Rorty, who sees science as part of a general human project to cope with the world around us, rather than a quest for deep understanding or 'truth'. Taking his cue from Darwin's view of human evolution, he argues that

> We do not differ in kind from the animals. All that distinguishes us is the ability to behave in more complex ways. The older, pre-Darwinian conception is that animals can't grasp 'how things really are', whereas humans can. From the post-Darwinian perspective there is no such thing as 'how things really are'. There are simply various descriptions of things, and we use the one which seems most likely

to achieve our purposes. We have a multiplicity of vocabularies because we have a multiplicity of purposes. As history goes along, new vocabularies develop as new purposes emerge. But none of these vocabularies or purposes will be more true to 'human nature' or to the 'intrinsic character of things' than any of the others, though the purposes served may get better.[10]

While these worries might be significant, they are by no means unavoidable conclusions of a Darwinian perspective on the origins of human intelligence. Our study of complexity in many different manifestations has taught us that there is rarely a smooth, steady increase in the consequences of similar changes in complexity: huge jumps occur when critical thresholds are reached. Our DNA may differ from that of chimpanzees by merely a couple of per cent, but the consequent intellectual complexities are light years ahead of a chimpanzee's. It may well be that this huge leap ahead of all other living creatures merely allows us to produce more extensive languages for the description of Nature. But the presumption that this is all we can do is hiding a belief that Nature is a work of bottomless complexity, whose surface we are always scratching. It may not be like that. There could be a bottom line. As we have already seen, there are a variety of options regarding the scope of Nature and the power of human discovery. So far, the Universe has proved to be far more intelligible than we had any reason to expect. Ironically, the most complicated thing we have encountered in the entire panorama of Nature, from the inner space of the elementary particles of matter to the outer space of distant galaxies, is what lies inside our heads.

Counting on words

No doubt our ancestors needed some rational skills to survive, but ... the human brain evolved more as a religious than a rational organ ... Rational science is a minority interest ... It is likely therefore that the first human brains evolved to impose symbolic meaning on the external world, and the scientific virus later infected a minority of their descendants, where it now flourishes in nerve circuits that originally evolved to carry other ideas.

NICHOLAS HUMPHREY[11]

The history of the mind delivers a pessimistic message to users. If we are engaged in a grand quest to understand the deepest logic the workings of the Universe, from the elementary particles of matter to the furthest reaches of intergalactic space, then we may find ourselves ill-equipped. There seems to be no reason why our early history need have equipped us to deal with tricky mathematical problems about quarks and black holes, no reason why we should possess the ability needed to visualize the most abstract mathematical structures, and no reason why our abilities to visualize should reach anywhere near the levels required to provide a description of the laws of Nature (if such there are). Yet, although all these statements might turn out to be true, we may be able to delve a little deeper into the nature of things. The optimistic pitch would be that the deep and difficult concepts needed to fathom the depths of Nature's structures are made up, step by step, from very simple concepts like counting, cause and effect, symmetry and pattern, asking yes/no questions, and so forth. After all, these simple building blocks of deep understanding do offer clear and simple advantages to their possessors when they join the evolutionary rat-race. The concepts of modern science may appear superficially very abstract and far removed from those required to survive in arduous ancient environments, but it is remarkable how simple are the basic concepts from which they are constructed.

It is often suspected that mathematics is so miraculously effective a tool in the unearthing of Nature's workings that it might well be that Nature *is* mathematical in some ultimate sense. Thus, our mathematical description of Nature is actually a process of discovery rather than one of invention. While this might well be the case, we still have to worry about how we have come to select certain types of mathematics to apply to the world, how we have set up our notations and concepts. Here, I believe, there are unsuspected links with our linguistic abilities.

The most impressive human ability is that of language. As we look at the population at large, we see that mathematical skills are very variable. Most schoolchildren find maths hard. No one seems to know it innately. We have to sit and learn its rules and structures. But language just came naturally. There are many individuals who have no mathematical or musical ability, but every able-bodied person speaks a language with remarkable sophistication. Moreover, they seem to manifest this ability at an early age in ways that argue for language being a genetically inherited ability. Listening and learning merely fix which native language will be spoken initially. All languages share enough of a deep logical structure for the brain to carry a built-in program for the acquisition of linguistic ability. Moreover, it is unnecessary for mental resources to be given over indefinitely to running this initial language-learning program. Once a language is learned in childhood the program can be switched off and resources channelled into other cognitive processes. This is why non-native languages are acquired differently. They require conscious learning effort, and the older you get the harder it becomes.

The counting systems that developed in ancient and traditional societies have many similarities.[12] They all involve the creation of symbols for quantities, and most of them group quantities into collections whose number is most commonly linked to the number of fingers (10), or fingers and toes (20), that we possess. More important are the concepts which these systems introduced to aid combining quantities—doing what we call 'adding up'—and creating handy notations for recording large numbers. Here we find several advanced cultures like those of the Babylonians and early Indians introducing the concept of 'place-value' notation, together with a symbol of zero. The place-value notation is very powerful. It is so familiar to us that it goes unnoticed. It means that the relative positions of symbols carry information about their numerical values. When we write '111' we understand it to mean one hundred plus one ten plus one (that is, one hundred and eleven), whereas for an ancient Egyptian it would have meant just one and one and one (that is, three). A place-value system leads inexorably to the need to register an empty slot; so '101' is one hundred, no tens and one. The invention of 'zero' by the Indian culture completed this highly efficient notation which is now completely universal.

This familiar structure for representing numbers has similarities with linguistic structures where the relative positions of words carry information. Languages usually have rules about the relative positions of adjectives and the nouns that they limit. We recognize the pattern of sentences so that we can substitute different verbs and nouns into particular slots in the structure. All this is too valuable and pervasive an ability not to have been used for some other simpler purpose. The similarity of different ancient counting systems and the notations that humans use to record numbers owes much to our instinct for language. The fact that quantities had to be talked about before they could be represented by symbols ensures that the way they are talked about influences the manner in which they are denoted by marks and symbols. At first, numbers seem to have been used as nouns. There was a word for three stones, another for three sticks, another for three fish.[13] The concept of threeness was always allied with the identities of things. This leads to a profusions of terms and symbols. But, think of numbers as adjectives and you can streamline your language with one word for 'three', which you place beside the word for any of the things whose number you want to describe.

The importance of this idea is that, if true, it shows how counting began by moving along a track that had been trodden by the development of language. Counting led ultimately to mathematics. Our mathematical notation, like our mathematical concepts, began as by-products of instinctive intelligence for other activities.

Modern art and the death of a culture

The fashionable oppish and poppish forms of non-art today bear as much resemblance to . . . exuberant creativity . . . as the noise of a premeditated fart bears to the trumpet voluntary of Purcell.

LEWIS MUMFORD[14]

Gunther Stent thought that his argument for the self-limiting nature of science was supported by looking at what had happened to the creative arts. Like many others of his generation and European cultural background he was mystified by the direction that the creative arts had taken. He noticed that many commentators (even some within the artistic community) thought that art was no longer 'real' art, but merely some form of spilt emotion. Taking a longer look at the situation, Stent tried to interpret the present state of affairs as the end result of an evolutionary process which has steadily relaxed the compositional constraints placed upon the artist. Over the centuries we have seen new materials and media appearing to enlarge how creativity may be expressed. At the same time the traditional restrictions on what may (or may not) be portrayed, and how it may be done, have been steadily eroded. As the constraints imposed by convention, technology, or individual preference have been relaxed, so the resulting structure is less formally patterned, closer to the random, and harder to distinguish from the work of others working under a similar freedom from constraint.

One of the characteristic features of appreciated music in all cultures is the way that it combines sequences of sounds to produce an optimum balance of surprise and predictability. Too much surprise and we have unengaging random noise; too much predictability and our minds are soon bored. Somewhere in between lies the happy medium. This intuition can be put on a firmer footing. Some years ago, two physicists at Berkeley, Richard Voss and John Clarke, discovered that human music has a characteristic spectral form.[15] The spectrum of a sequence of sounds is a way of gauging how the sound intensity is distributed over different frequencies. What Voss and Clarke discovered was that all the musical forms they examined had a characteristic spectral form, called '$1/f$ noise' (pronounced 'one-over-eff noise') by engineers, which is precisely the optimal balance between unpredictability and predictability: there are correlations over all time intervals in the sound sequence.

We can add something to this characterization of music by applying it to the style of the composition. When a musical composition is in a style that is highly constrained by its rules of composition and performance, it will be far more predictable than if its style is free from constraints. The listener does not receive very much new information, over and above that present to establish the stylistic framework, from listening to the music. Conversely, if the style has too few constraints, the unpredictabilities in the sequence of sounds can be too great. An instant appreciation of the weak probabilistic patterns of sounds will be hard to make, and the result will be perceived as less attractive than the optimal $1/f$ spectral pattern.

Stent argued that music must evolve in the direction of greater stylistic freedom. Because of the cumulative nature of previously created works in each genre and the growing sophistication of the listeners' appreciation, it is the only place left to go. Starting with the maximal rigidity of rhythmic drumming in ancient times, music has exhausted the scope of each level of constraint for its listeners, before relaxing them and moving down to a new level of freedom of expression. At each stage, from ancient to medieval, renaissance baroque, romantic, to the atonal and modern periods, evolution has proceeded down a staircase of ever-loosening constraints, the next step down provoked by the exhaustion of the previous level's repertoire of novel patterns.

This evolution is one of increasing sophistication in information-processing with time. The invention of musical notations and new media for recording and replaying music privately greatly accelerated the sophistication process, giving many more ways in which to develop away from the constraints of average tastes. The culmination of this evolutionary process in the 1960s saw composers like John Cage relinquish all constraints, leaving the listeners to create what they would from what they heard: an acoustic version of the Rorschach inkblot test. Instead of communicating satisfying patterns, they sought to evoke transcendental experiences. Their music does not invite interpretation as a correlated temporal sequence of sounds; it just *is*. Distinguishing music from noise depends entirely on context; it is sometimes impossible, and even undesirable.

Other creative activities, like architecture, poetry, painting, and sculpture have all displayed similar trends away from constraint. Stent's suspicion was that they were all quite close to reaching the asymptote of their stylistic evolution: a final structureless state that required purely subjective responses. The future predicted by the musicologist Leonard Meyer was that

> the coming epoch (if, indeed we are not already in it) will be a period of *stylistic stasis*, a period characterized not by the linear, accumulative development of a single fundamental style, but by the coexistence of a multiplicity of quite different styles in a fluctuating and dynamic steady-state.[16]

An alternative to this picture of decay and dissolution is that of cyclic evolution, in which the styles of the past are resurrected and reused. If an art form, such as popular music, displays steady technical progress in sound production and processing, then this recycling of old material can be very tempting. It is certainly very common.

This pessimistic picture of artistic evolution, which has focused for simplicity on musical development, is one of diminishing returns in the face of successful exploration of each level of constrained creative expression. To escape from its clutches, individual creativity has to assert itself. Diversity has to be fostered. This should give us pause for thought, because so many of our technical and social developments do the opposite. We view greater collaboration and easier connections between people and organizations as a measure of progress. But where there is wide collaboration in artistic creation there is the danger that diversity will die. We shall return to this question later in this chapter, after we have explored the impact of artificial minds upon our own.

Complexity matching: climbing Mount Improbable

The brain is a three-pound mass you can hold in your hand that can conceive of a universe a hundred-billion light years across.

MARIAN DIAMOND

Our quest to understand the structure of the Universe and the rules that govern it may succeed or fail. There is no guarantee. Its outcome depends upon a close match between the complexity of our minds and the complexity of the Universe. Since such a match seems unlikely, we might have expected that our minds would either fall far short of, or far outstrip, what is required for understanding the Universe. A close match smacks of a very peculiar coincidence.

The fact that our minds have evolved by an adaptive process means that our level of mental sophistication has been driven by particular problems that the real world has set. An ability to solve them better than some rival confers an advantage and selection begins. If this were the whole story, then we could quickly conclude that our minds have encountered only a tiny part of the natural structures of Nature, so they must fall far short of the level required to unravel the entire puzzle. But things are not so simple. Our minds seem to be far more powerful that are required for mere survival. We possess many skills which are not just slightly better than those of other living things. We are streets ahead of them; ahead by so much that we have largely now transcended evolution by natural selection. We can use our imaginations to simulate the results of our actions. We do not have to learn only from our mistakes, passing on information genetically from generation to generation; we can pass on information by word of mouth, over the airwaves, on the Internet, or in print. This information can influence any member of the species that hears it. The time is takes to convey information is now very short and its influence extremely wide.

Despite this remarkable state of affairs, which seems to have emerged quite suddenly at some point in our history, and for reasons that are still not clear,[17] we are obviously limited. We have constructed electronic computers which can compute far more rapidly and reliably than we can. We can begin to sense how some future generation of machines might overshadow us in many other respects as well. When we try to extrapolate human science into the far future it seems clear that artificial forms of intelligence are going to develop far more rapidly than our innate mental abilities. Indeed, the latter might even decline in certain respects because we no longer need to perform many of the feats of mental dexterity that were once everyday requirements. Quick mental arithmetic is fast fading as a skill among young people: calculators make it of little value now. Thirty years ago, the child who wanted to take a weekend job serving in a shop would be judged on his or her ability to do fast and accurate mental arithmetic. Now all the goods are swiped past a scanner and the bill is printed out—even the change is computed for you.

Seen in this way, the development of artificially intelligent systems and of more powerful computer systems looks like another decisive step in the evolutionary process, not unlike the evolution of language. Human language enabled us to develop large amounts of interaction between individuals, to pool information and experience, so as to learn faster than by living and learning in isolation. The evolution of human civilizations witnesses to the constant search for better means of communication. In the technological era, our greatest discoveries have been the means for conveying information over large distances virtually instantaneously. Radio waves, telephones, optical fibres, the Internet, satellite communication systems: all these have enabled more minds to apply themselves to more problems in a shorter period of time than ever before. Already, we can foresee a not-too-distant future in which all computers on the planet will be linked simply and inexpensively into a global network. These developments may produce a computer future different from what many futurologists were predicting not too many years ago. In 1943, Thomas Watson, the chairman of IBM said, 'I think there is a world market for maybe five computers.' Even in 1977, Ken Olsen, the founding president of the Digital Equipment Corporation (DEC), was of the opinion that 'there is no reason for any individual to have a computer in their home.'

Everyone expected that computers would just keep getting bigger and more and more powerful (and more and more expensive). But this is not what happened at all. Computers got smaller and smaller (and cheaper and cheaper). More and more people owned them, and their effectiveness developed most impressively by linking them together into huge networks. Similarly, there is little to be gained by any further evolution of the intellectual capabilities of single brains. When they have reached a level of sophistication deep enough to appreciate and effect collaboration with many other minds, then the benefits of so doing completely outstrip those obtained by increasing the power of individual minds. In any complex system it is not so much the size of the components that is of primary importance, but the number of interconnections between them. This number grows very quickly as the number of connection points increases. The number of possible links between six connection points is enormously greater than the number of possible connections in two separate collections of three points, as Fig. 4.3 illustrates.

So, even without raising speculative (but not implausible) expectations about our abilities to find faster and cheaper means to carry out computations, we can

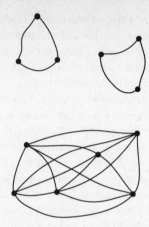

Fig. 4.3 Connectivities. Two networks of three points have two connections emanating from each point, giving a total of six. If these six points are made into a single network, then each point has five connections, giving a total of 15.

foresee evolution towards a future in which the networking of computers can stimulate and solve problems of vast size and complexity. The intrinsic complexity of the programs, and the self-programming abilities of the computers that will train themselves to perform these feats, will gradually approach the complexity of the physical problems they seek to solve. As they do so we shall have to ask what is meant by saying that they have 'solved' these problems. If we can introduce a number of inputs and rules of computation, then we shall be content that, no matter how complex the path to the output, if it agrees with what is seen, we have understood the solution to the problem posed.

We would probably be happy with this judgement if the computer were doing a vast number of simple things for us, the outcome of each one being something that we could easily visualize. But if the computer were to perform a lengthy process of computation involving individual steps of such complexity that we could not fully envisage their outcomes, then we would begin to worry about what our 'understanding' really amounted to. A full simulation of a complicated natural phenomenon would involve a program of complexity approaching that of the thing being studied. It is like having a full-scale map, as large as the territory it describes: extraordinarily accurate; but not so useful, and awfully tricky to fold up.

These speculations take us closer to understanding how our minds might fall short of what is needed to understand the Universe—or even quite small parts of it. We might face the problem that our minds cannot grasp certain concepts at the bedrock of reality. We are straitjacketed by our evolutionary history and our innate linguistic abilities (which do not vary much from person to person), and it might be judged a very fortunate state of affairs if we just happened to be smart enough to accommodate all the concepts required to formulate a correct Theory of Everything, for instance. To paraphrase J.B.S. Haldane, the Universe may not be only stranger than we imagine, it may be stranger than we *can* imagine.

Our mental abilities were laid down long before the notions of modern physics emerged in our minds. Our only hope of dealing with such an eventuality is to hope that abstract concepts will always be divisible into collections of simpler ideas. While it is easy just to hope that we shall always be able to cope with new ideas, it has become clear that the mathematical structures employed at the frontiers of fundamental physics theories are becoming less and less accessible—even to physicists. The number of individuals on the planet with the ability to understand the mathematical structure of superstring theory is relatively small. It would not require mathematical structures to be very much more sophisticated in size and complexity for that number to shrink to zero.

At present it is fashionable to believe that there is a 'bottom' line in fundamental physics: a basic collection of indivisible entitles obeying a small number of mathematical rules in terms of which everything else can in principle be described. but the world may not be like this. Like a sequence of Russian dolls, there may exist an unending sequence of levels of complexity, with very little (if any) evidence of the next level down displayed by each of them. If this is the case, then we are as far from knowing the whole story as we have ever been, or ever will be.

More constraining than these conceptual limitations is likely to be our inability to visualize and coordinate large complex structures. Fast computers will take us into a new realm of hard problems. So far, we have spent thousands of years building up our knowledge of the simple structures of Nature and logic. Simple structures are those which can be built up from basic components in a small number of steps in an elementary way. 'Small' and 'elementary' here mean by using pencil and paper or other simple calculating devices. It is rare for us to take a strong interest in problems whose shortest solution is of enormous length. The mathematics that we know and use lives in the realm of short truths. Perhaps the deepest truths are the longest. Only fast computers can take us into the realm of long deep truths. It is like climbing high mountains. The unaided rambler cannot ascend very far; the climber aided by ropes and tools can go much further; and the rarefied heights require yet more artificial aids: oxygen, special clothing, and food. But there could be a mountain so high that the provisions and equipment needed for the ascent are just too much to be carried.

Intractability

And the children of Israel were fruitful, and increased abundantly, and multiplied, and waxed exceeding mighty; and the land was filled with them.

EXODUS[18]

The question of how hard problems might prove to be seems rather subjective. What one person finds easy, another might find hard. Fortunately, there is more to be said. Computer scientists have spent a good deal of the past 25 years devising a classification of the degree of difficulty of problems that can be attacked using any computer. This has led us to distinguish between tasks which are impossible in principle, of which we shall hear more in Chapter 7, and tasks which are 'practically impossible'. By 'practically impossible' we mean that it would take a prohibitively long time to solve them by the fastest program that could be written. Such problems are generally referred to as *intractable*. A typical example is that of 'the travelling salesman'. Suppose a salesman must visit N different cities; given the list of cities and their distances apart, find the optimal route for him to take which minimizes the total distance he has to travel. When the number of cities, N, on the itinerary is small, this is easy to solve by trying the alternatives. In Fig. 4.4, a simple six-city example is shown, together with the shortest route. But as N grows very large the time that is taken to do the checking grows very rapidly with the size of N. In general, no recipe is known which will supply the salesman's optimal routing. In Fig. 4.5, we show the largest itinerary for which the problem has been solved. This solution is commercially important. It is not a route for a salesman to follow as he goes from city to city; rather, it is a plan to wire a circuit board for a computer in a way that minimizes the time and energy used. Since the number of circuit boards produced is so large, any improvement in production time converts itself into a large financial saving in the long-term production process.

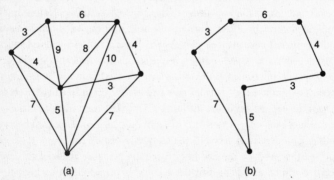

(a)   (b)

Fig. 4.4 (i) A simple six-city Travelling Salesman Problem, showing the distances of trips between all pairs of cities. (ii) The shortest route which visits all six cities.

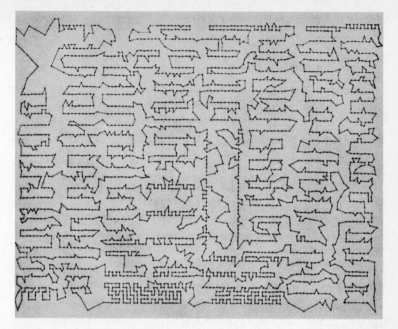

Fig. 4.5 The largest solved Travelling Salesman Problem, with a tour of 3038 sites, found by David Applegate, Robert Bixby, Vadek Chvátal, and William Cook, in 1990. This is a printed circuit board showing the shortest route for a robot to visit all the holes and make electronic connections. The solution took one and a half years of computer time to find! The previous record holders solved a 532 site tour in 27 hours of computer time. Notice how the strategy adopted by the algorithm varies from one region of the problem to another, depending upon the pattern of inter-node spacings.[19]

There are many other combinatorial assignment and routing problems with a similar flavour: for example, assigning teachers to classes in a school timetable so that there are no clashes, or finding the smallest volume into which a collection of objects of different shapes and sizes can be fitted.

Let us consider two more example of puzzles which teach us something more about how hard we have to work to solve soluble problems. The first is the monkey puzzle (Fig. 4.6). This is the poor man's Rubik's Cube. It consists a collection of nine square cards on which four halves of coloured monkeys are drawn. The aim of the game is to arrange the nine cards so that they match the correct parts and colours of the monkey's bodies wherever the cards meet. A computer algorithm would have to work its way through $9 \times 8 \times 7 \times 6 \times 5 \times 4 \times 3 \times 2 \times 1 = 362,880$ configurations to find the solution. We call this quantity 'factorial nine', and represent it by the symbol 9! The factorial operation grows very rapidly. The quantity 36! contains 41 digits, and a

Fig. 4.6 A Monkey Puzzle with nine pieces. Nine cards are printed with upper and lower halves of coloured monkeys. The aim is to arrange the 3 × 3 array of cards into a square so that whole animals are formed with a single colour wherever the edges meet.[20]

computer trying this number of options at the rate of one million per second would take more than eleven billion billion billion years to work its way through all the 36! arrangements. This is certainly a 'practical' impossibility.

Our second example is the ancient problem of the Towers of Hanoi. The problem was first introduced to mathematicians by Walter Rouse Ball, an English connoisseur of mathematical problems, who attributed it to the following legend:

> In the great temple in Benares . . . beneath the dome which marks the centre of the world, rests a brass plate in which are fixed three diamond needles, each a cubit high and as thick as the body of a bee. On one of these needles, at the creation, God placed 64 disks of pure gold, the largest disk rest resting on the brass plate, and the others getting smaller and smaller up to the top one. This is the Tower of Bramah. Day and night unceasingly the priests transfer the disks from one diamond needle to the other . . . [such that only one disk may be moved at a time and no disk must be placed on top of a smaller one] . . . When the 64 disks shall have been thus transferred, from the needle on which at the creation God placed them, to one of the other needles, then tower, temple and Brahmins will crumble into dust, and with a thunderclap the world will vanish.[21]

The set-up is pictured in Fig. 4.7. The ingenious creator of this task clearly intended to keep the priests busy for a long time. In fact, if there are N disks to be shifted between three needles according to the rules of 'one disk at a time' and 'no disk on top of a smaller one', then the transfers required cannot be completed in less than $2^N - 1$ moves.[22] So, in the case posed, where $N = 64$, even

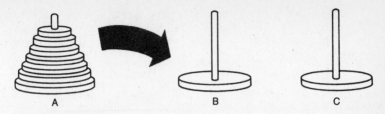

Fig. 4.7 The Towers of Hanoi Problem. The aim is to move all the disks from peg A to peg B, using peg C, in accord with the given rule.

making very rapid progress with one move per second would take 412 billion years! For comparison, the Universe appears to have been expanding for less than 10 billion years. Even if we were to miniaturize the operation and use a computer operating at a billion moves per second to carry out the transfers, it wouldn't help much: it would still take more than 400 years.

Both these two problems are examples in which the algorithm needed for their solution grows more rapidly than any mathematical power of N (that is, faster than N^k for any number k) as N, the number of ingredients to be dealt with, increases. Problems which, on the other hand, grow in difficulty as some power of N fall within the class known as **P**, for polynomial, to emphasize that they can be solved in a time that depends only on some mathematical power of their size, N. One might regard these as 'easy' problems. Most of the things that we employ computers or calculators to do, such as adding up lists of numbers or addressing envelopes for a bulk mail-shot from an address directory, are of this type. By contrast, neither the Monkey Puzzle nor the Tower of Hanoi puzzles are of class **P**, because 2^N and $N!$ both grow faster than N^k for *any* value of k as N grows very large.

Figure 4.8 gives some indication of just how fast some of the numbers we are talking about grow as N gets large.

Harder problems are classified as **NP**, for non-deterministic polynomial, if their solutions cannot be verified in polynomial time.[23] The steepest curves in Fig. 4.8 characterise the growth of **NP** problems. Note that in all these examples we have focused solely on the running time of programs; we have said nothing about the memory storage space required to handle the information generated by the computations. Even a modest computation that grows as 2^N as the number of steps, N, grows, will very soon require the entire visible universe to store intermediate computational information, even if it writes one bit of information on a single proton, because there are only about 10^{79} protons available.

In practice, it is very difficult to prove that any given problem cannot be solved in polynomial time (the examples we have given are very simple problems). At present, there are no more than about a thousand problems

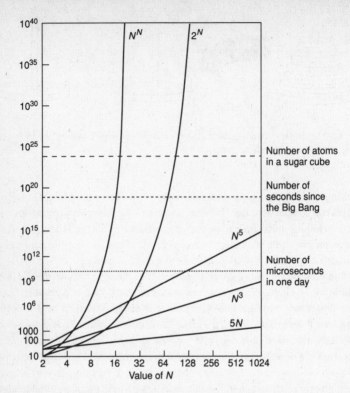

Fig. 4.8 The rate of growth of the quantities $5N$, N^3, N^5, 2^N, and N^N as N increases, together with some other large numbers for reference.

which are suspected to be **NP**. One of the great unsolved problems of modern mathematics is to determine whether no **NP** problem is a **P** problem; that is, to discover whether you can find a solution in polynomial time if you only need polynomial time to verify that it is true.

Remarkably, virtually all known **NP** problems have been found to be very similar to each other in innate difficulty. In 1970, Stephen Cook, then a graduate student at the University of California at Berkeley, made a great discovery; he showed that all these **NP** problems can be related to a single problem in logic by a general transformation procedure that can be performed in polynomial time.[24] What this means is that if anyone finds a way to solve *any* **NP** problem quickly, then it would be possible to solve *all* the other **NP** problems quickly.

Intractability has created a perplexing problem in molecular biology. Every living organism contains proteins, formed from chains of amino acids, like the beads on a necklace. When the amino acids appear in the correct order, the

DNA provides the information which specifies how the protein will fold up to form a complicated geometrical structure in three dimensions. This process is called *protein folding*. An example[25] is shown in Fig. 4.9.

The problem that molecular biologists would like to solve is this: if we start with a given linear chain of amino acids, what is the particular three-dimensional configuration into which it will fold? Remarkably, we see chains of several thousand amino acids folding into their final pattern in about one second. We think that the final shape is one that minimizes the energy required to support its structure, but when we try to program a computer to fold a protein, it seems to be impossible. If there were just 100 amino acids in the chain it appears to require more than 10^{27} years to effect the folding! To confuse us even further, in 1993 the mathematician Aviezri Fraenkel showed that our formulation of the protein folding problem is NP-complete, and hence as intractable as the Travelling Salesman Problem.[26] Other studies, using pro-grammed heuristic reasoning, which extends the scope of computational studies, have been performed by George Rose and his collaborators.[27] They fed in a variety of rules of thumb to help the system to fold, telling the computer various simple chemical principles so that it did not have to search through every possibility. This technique can predict much of the folded protein struc-ture. It appears either that the process of natural selection has built living systems out of proteins that fold especially easily, or that the folding problem does not need to be solved by Nature with perfect accuracy. It may be enough to do it approximately the same each time. Proteins for which small mistakes were lethal would not survive. Others have suggested that Nature may be carrying out its computations in a massively parallel fashion, like a 'quantum' computer, rather than in series as we do.[28]

The intractability of problems which are very simple to pose, like finding the factors of a very large number, is currently so great that they form the basis of modern forms of encryption.[29] All you need to do to break the code is be able to find the two huge prime numbers that may have been multiplied together to form a composite number with, say, a hundred digits. This is something that takes so long for the computer to do that the code is secure in practice. The largest number ever factorized into two prime factors is 167 digits long. The two factors are 80 and 87 digits long. This feat took Samuel Wagstaff and his colleagues at the University of Indiana 100,000 hours of computer processing.[30] These numbers are shown in Fig. 4.10.

These examples give us a glimpse of what tractability means when we use computers (or any other means of calculation) to attack hard problems. However advanced our computer technology becomes, we shall be faced with trying to carry out tasks that are astronomically long (this, incidentally, is an understatement; there are only about 10^{11} stars in our galaxy, and about the

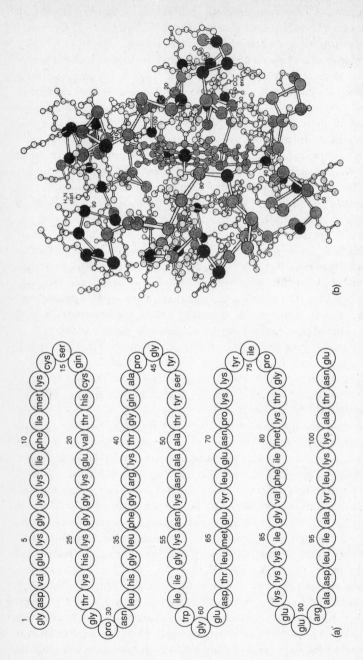

Fig. 4.9 A protein folding problem: (a) The protein *cytochrome c* consists of a string of 104 amino acids which, in a few seconds, folds itself into (b) a stable three-dimensional structure.

$$(3^{349}-1)/2$$

$$=$$

163790195580536623921741301546704495839239656848327040249837817092396946863513212041565096492260805419718247075557971445689690738777729730388837174490306288873792840 41

$$=$$

94042850889984510998289152320438541798532018021653956283741193211654025280185459

$$\times$$

17416549374087525646474638899948053399094433426684968705461152492287884070820660886 0499

Fig. 4.10 The largest known factorization of a number into two prime numbers, found in 1997 by Samuel Wagstaff and colleagues at the University of Indiana.

same number of galaxies in the visible Universe). We shall find ourselves faced with breaking a code that the Universe has used to encode information. The laws of Nature may transform the initial state of the Universe into a complex future state in such a way that even though we know both the laws that effect that change, and the present state, we are unable to invert the process to deduce the initial state because the computation is intractable.

This problem began as a failing of the human mind. We are well aware that there are calculations which, while simple in principle, require so long to complete that they are impossible in practice for an unaided human calculator (try writing down the first two billion numbers using pencil and paper). The development of computers does not save us from this problem. Even if we network huge numbers of computers together, we face the possibility that a full understanding of complex phenomena in many natural situations will take more time than we can ever spend on their analysis. Life is short; calculations can be very, very long.

The frontier spirit

Not even New Guineans can find enough wild foods to survive in the mountains . . .
This, until the advent of planes made airdrops possible, meant that all New Guinea
expeditions that penetrated more than seven day's walk from the coast did so by
having teams of porters going back and forth.

JARED DIAMOND[31]

A future prospect of interconnected intelligences pooling information and processing capability is an important consideration when we confront another worry about the ability of humans to cope with ever-expanding knowledge. If we simply scale up our present rate of acquiring knowledge, then we seem to be heading for a crisis of human capacity. At present, it takes about six years of secondary schooling, followed by three years of university study, before science students are equipped to start understanding what is going on at one of the frontiers of a mathematical science. It then usually takes two or three years before they are able to make unaided contributions to knowledge. This educational path is not optimized for scientific research of course: it must accommodate all sorts of people. Clearly, it takes considerable time and effort to reach one of the frontiers of human understanding. Most students never reach one at all. As our knowledge deepens and widens, so it will take longer to reach a frontier. This situation can be combated only by increased specialization, so that a progressively smaller part of the frontier is aimed at, or by lengthening the period of training and apprenticeship. Neither option is entirely satisfactory. Increased specialization fragments our understanding of the Universe. Increased periods of preliminary training are likely to put off many creative individuals from embarking upon such a long path with no sure outcome. After all, by the time you discover that you are not a successful researcher, it may be too late to enter many other professions. More serious still, is the possibility that the early creative period of a scientist's life will be passed by the time he or she has digested what is known and arrived at the research frontier.

The 'growth and fragmentation of knowledge' problem is one that we already face. Look at the telephone on your desk. How many people know everything needed to make a working telephone? There are acoustics, electronics, plastic design, economics, advertising, accountancy, metalworking, materials science, production engineering, chemistry, packaging, and on and on. No one person knows all these things to the level required. It is the same for just about all the technical devices that populate our homes: computers, electric typewriters, microwave ovens, hi-fi equipment, and television sets. These things are fruits of our collective activities. We have learned to coordinate different specialists so that the output from their individual contributions is more than the sum of the parts. If we required one person to become proficient in every aspect of technology and business that is required to manufacture and market a telephone, then telephones would still be rather rare and primitive. Science is just the same. It has become very much a collective exercise. Theorists rarely perform experiments. Important experiments and observations are carried out by large teams of individuals with many different talents: managers, engineers, physicists, statisticians, computer scientists, and electronics specialists. Again, the rapid progress of science reflects the efficiency of these collaborations. The most expensive scientific project, aimed at pushing back the frontiers of astronomy, particle physics, or molecular biology are already so elaborate and expensive that they require international collaboration.

This state of affairs raises the prospect that we might be able to avoid the problem of receding scientific frontiers, together with that of the increasing knowledge required to reach them, by extending the collectivization of some of our activity: by teamwork in which the teams might gradually become dominated by computers rather than by people. This picture of a future strategy to keep the frontiers within reach for as long as possible give grounds for optimism. It sees a scientific future of greater and greater international collaboration. There seem to be many potential benefits. Scientific collaborations are already a model for civilized human interactions. Large international projects, like those in experimental physics at CERN, are well managed and free of many of the prejudices and irrationalities that characterize other human efforts to internationalize affairs. Perhaps this is because they involve a relatively small number of individuals with common aims. Or perhaps there is something about the scientific habit of mind that rubs off on the entire activity. On this view, a future of ever greater collaboration seems the answer to all our problems of fragmentation and over-specialization. But is the prospect entirely good news?

The end of diversity

Only connect . . .

E.M. FORSTER

The future of computers and forms of information processing is a chancy business to predict. Progress is swift and unsuspected by innocent outsiders. But if we look at the direction of the progress that has occurred recently, we see clear pointers for the future. The game is not, as prophesied by the pundits thirty years ago, to build bigger and bigger single machines capable of processing and storing more and more bits of information. Instead, computers are getting smaller and cheaper. What is growing is the capability for networking: creating the power of a huge interconnected computer through a web of connections between an ever-growing number of small devices. Nature got there first. This is a pattern that has evolved within the human brain and in other complex organizations in the natural world. Some, like the intricate organization of the ant colony, call into question our prejudices about what constitutes a species or a living thing. By evolving a colony whose members have different capabilities and functions, a certain overall efficiency has been optimized. As computer and cognitive scientists try to unravel the workings of the brain and recreate some of its tricks and successes, we can expect to learn of new ways in which linking channels of information transfer together in parallel can work faster and better.

Let us take a step sideways to consider the wider implications of this general trend towards connection. The most prominent current example is the Internet and the associated World-Wide Web. It has changed the face of many human activities and professions. Many of these changes are positive. Scientists working in economically underdeveloped countries now have ways of keeping instantly abreast of developments in science and medicine. In the pre-electronic age they would have needed to subscribe to many fiendishly expensive journals in paper form. New discoveries can now reach a global audience almost instantaneously. Research groups can save money on telephone calls, faxes, and postage. Computer conferencing, although not yet common, offers the chance to spend less on travel. The efficiency of the scientific process is undoubtedly improved. In my own field, progress was significantly slowed in the 1950s and early 1960s by the fact that two of the world's foremost research groups at Princeton and Bell Labs were unaware of earlier theoretical predictions about the existence of the cosmic microwave background radiation in the Universe. This could not so easily happen now. Communications are faster and much more wide-ranging.

There are many other advantages of the growing interconnectivity of all forms of human knowledge for educators. But what of the disadvantages? They are rarely discussed. Yet, even if they are small, if they do exist, then they might come to dominate over the advantages in the long run. While they may not bring the whole human quest for further knowledge crashing to a stop as a result of some uncontrollable computer virus destroying everything, or grinding to a standstill through overuse, these adverse factors might well move human progress along certain tracks. The very success of the system might entrain the way we think. Unnoticed, it might dictate the types of question that get asked and the sort of answers that are found.

One of the most noticeable effects of global connectionism among scientists is the way in which it facilitates and encourages of a large international collaborations and groupings. This is not unconnected to wider (and often highly controversial) political aspirations among the economically developed nations. The Internet makes research blind to proximity and scale. It is easy to draw together the plans of groups in different countries. The European Union has tried to take advantage of this by encouraging the creation of networks of researchers in different European countries, with a view to using the scientifically advanced members to train and aid the less developed. One adverse result of this trend, which might multiply when extrapolated into the far future, is a reduction in diversity of view. Gradually, each subject area tends to become a single research group. In the past, it was common for there to exist separate groups in different parts of the world, who would develop their own approaches to a problem before detailed interaction occurred with other groups. This has changed. Single, central paradigms are now strongly reinforced, and young researchers become increasingly involved in detailed elaborations of them. In effect, there is increased specialization and focus. This affects not only the subject matter under study, but the style of its study. Interpersonal contact is reduced, and contact with books and printed journals is minimized. Paradoxically, these trends have common consequences: they remove the chance of discovering new things by chance. Scientific journals publish a mixed bag of articles on a subject which might be as broad as 'astronomy' or 'mathematical physics'. If you need to search for an article that you remember appeared in the journal *Societ Astronomy*, some time in the mid-1980s, then you will need to search the subject or author index of several volumes to track it down. Invariably, in my experience, that search process leads you to discover other interesting things: some directly relevant to your current interests, others, striking, but off-beat, to be filed away in your memory for some future application. By contrast, a computer archive would allow you to find what you were looking for without any risk of serendipitous discovery. The human consequences are not dissimilar. If you can obtain information about subjects other than your speciality from the computer network, you will be less likely to seek out others to ask them about the things you want to know. Accessing a computer is so easy, whereas seeking out individuals for discussion requires a positive effort that is ever more likely to be eschewed by busy people.

These two simple examples are not supposed to exhaust all the negative consequences of global computer connections. They show how the medium can produce subtle changes in the way in which interactions between minds take place, what sorts of questions are addressed by the scientific enterprise, and the likelihood of making unexpected discoveries by chance. At present these influences and biases might be quite small. Over a very long period of time their effects might well become dramatic and irreversible.

Does science always bring about its own demise?

> *The most merciful thing in the world . . . is the inability of the human mind to correlate all its contents . . . The sciences, each straining in its own direction, have hitherto harmed us little; but some day the piecing together of dissociated knowledge will open up such terrifying vistas of reality . . . that we shall either go mad from the revelation or flee from the deadly light into the peace and safety of a new dark age.*
>
> H.P. LOVECRAFT

Knowledge carries dangers. It can be used or abused. It can lead to disaster by intent, or by accident. As scientific knowledge and technical expertise have accelerated, we have come to appreciate its dangers. In the second half of the twentieth century, humanity reached a critical state in its development. For the first time it had the means to initiate a global disaster of sufficient magnitude to extinguish itself. We know that a few wrong moves by politicians and military leaders in the 1950s and 1960s might have unleashed a nuclear holocaust. Today, new diseases against which we have no natural resistance threaten us; irresponsible industrialization may have set in motion irreversible climatic changes that will make our planet a less comfortable, and ultimately an uninhabitable, place. The pressure on natural food sources and raw materials is steadily increasing. It is not hard to envisage a future in which there is no future. The growth of technology and the political trend towards deregulation tends to put increasingly powerful devices and processes in more and more unregulated hands. The total reliance upon computer systems for storing and controlling vital information puts it increasingly at risk from theft, manipulation, or corruption. As these sensitive systems become more complicated, so they become more vulnerable.

I have deliberately painted a somewhat pessimistic picture. I want to show that it is quite reasonable to speculate that technological societies never progress far beyond the point at which they have the means to destroy themselves. If nuclear weapons had been discovered just a little sooner in Germany or the USA, they might have been used on many occasions during the Second World War. It is also important to recognize the close link between the technologies needed to further research into fundamental physics and the industry of weapons of mass destruction. Any extraterrestrial civilization that investigates the former will have the means of producing the latter. The same is true of biological progress. Biological weapons have been produced in huge quantities, although used only by small powers in local wars. It could easily have been otherwise. Again, a certain degree of fundamental progress in the understanding of biochemistry, of a sort that would be inspired by laudable medical curiosity, necessarily possess a potential dark side of application. It is this Janus-like aspect of many parts of science which leads one to take seriously the prospect that scientific cultures like our own inevitably contain within themselves the seeds of their own destruction.

If true, this analysis would mean that we are unlikely to be troubled by our conceptual or technical limitations. On the contrary, it is their relative *absence* that will be the end of us. Our instinctive desire for progress and discovery will stop us from reversing the tides in our affairs. Our democratic leanings will prevent us from regulating the activities of organizations. Our bias towards short-term advantage, rather than ultra-long-term planning, will prevent us from staving off disasters that are slow and gradual, worsening imperceptibly during a typical human lifetime.

If catastrophe is not to engulf us, we shall have to overcome the problems we have just discussed, together with many others that have yet to rear their heads. Even if our ingenuity and collective responsibility are up to the challenges they present, the demands of overcoming them will so dominate the direction in which our resources are channelled, and the problems upon which scientists concentrate their minds, that our culture will be significantly altered. Knowledge for its own sake may become an increasingly profligate investment of human resources. Just as leading scientists were once enrolled into teams to devise new techniques for attack and defence in times of war, so scientists of the future may find their work redirected by conscience, or by necessity, in order to solve life-threatening global problems.

Death and the death of science

Nature, Mr. Alnutt, is what we were put here to rise above.
KATHARINE HEPBURN[32]

The growth of knowledge shares many features with living things. Ideas multiply and mutate, successful ideas survive to pass on information to the future. But living things are not immune from extinction, any more than their individual careers are immortal. We have considered some ways in which science might continue, significantly slow down, or even grind effectively to a halt. In practice, the limits of science are dominated by the limitations of scientists rather than by fundamental restrictions upon what can be known. Thus, we can envisage a future in which human fallibility becomes more and more significant. When mistakes are made and wrong conclusions drawn from evidence, they could be categorized by the amount of time it takes for the error to be corrected. Some are corrected almost immediately; others, like 'cold fusion' take a little longer; others, like Aristotle's theory of motion, might persist for more than a thousand years. We are used to living in an environment where scientific successes outnumber mistakes, so that there is a gradual increase in our fund of tested and effective information about the Universe. As the seams of easily accessible knowledge are mined out we shall have to dig deeper for new truths. These truths will be harder to find, more susceptible to erroneous or incomplete formulations, and therefore less reliable as bases for technological innovations. We can easily envisage progress into an era where mistaken deductions become the rule rather than the exception and scientific knowledge becomes unreliable.

This pattern of scientific demise has much in common with some theories of human ageing and death. In our prime, DNA copying errors can be corrected very rapidly by the error-correction procedures appended to our genetic coding. As we age, these correction devices become less and less effective, and copying errors can accumulate to lethal levels. Applied to the growth of reliable knowledge, this is a scenario of unspectacular demise: a world that ends not with a bang but a whimper, overcome by a rising tide of uncertainties and minor mistakes. At first, they just reduce efficiency; but eventually their effects are paralysing and scarce resources are squandered. We have already raised the question of whether human fallibility will result in our quest for a full under-standing of our Universe falling short of its objectives. We also have to consider whether that same fallibility might halt the increase of reliable knowledge so far short of its ultimate goal that our continued existence would be threatened.

The psychology of limits

> *The difference between a conjuror and a psychologist is that one pulls rabbits out of a hat while the other pulls habits out of a rat.*

ANONYMOUS

We have been discussing how the limits of the human mind might influence the ultimate achievements of science. We should not forget that a study such as this is itself a product of the mind's subtlety and we can legitimately inquire about the bias that this introduces. Is there some psychological component to pro-nouncements about the limits of science? Does a certain experience of science itself induce a particular attitude towards the future progress of science. Let us consider some possible correlations.

If a scientist is in the prime of life for invention, and new results are coming thick and fast, then he will not want this golden period to end and he will believe that it can't and won't. This view is likely to be reinforced if the new phase of rapid progress in which he is playing a central role is the immediate result of the overthrow of an old theory. Conversely, if a scientist's creative powers are failing, he may find that the most comforting rationalization for their diminishing effectiveness is to believe that the field as a whole is becoming less fruitful; that its harvest of new discoveries is steadily falling, and may one day dry up altogether. It is easy to imagine your own pattern of life is a template for the development of science as a whole. Curiously, this tendency need not be correlated to the level of creative activity in science; indeed, it may be negatively correlated with it. The once-active researcher may feel the reality of his own waning powers more strongly in a period when the subject is being enthu-siastically pushed forward by others. There is a tendency for former leaders of a subject to react to this state of affairs by becoming strongly opposed (often for philosophical reasons) to the whole direction of this advance. If they once made important advances by swimming against the tide of scientific opinion, they tend always to want to do the same, almost regardless of the strength of the evidence.

We might wonder whether a consideration of limits of science is an activity only for older scientists. Young people need to be consumed by the desire to solve soluble problems, if they are to be successful scientists. But the younger that scientists are, the closer they are to their apprenticeship of systematic instruction as students. During this period, they see only soluble problems; the practicalities of educating them, of examining their knowledge by some quantitative measure, of studying problems that primarily need pencil and paper for their solution and can be completed in manageable periods of time, create a bias. This bias in our education of scientists, particularly those engaged in the mathematical sciences, leads to the unconscious assumption that all problems are soluble. When students come to a university to study mathematics or physics they tend to believe that all integrals and differential equations are exactly soluble. Of course, they are not, but the students have only ever seen ones that are. Those who begin their careers as research students have to learn many new advanced topics. How they do that is quite important because it will impose limitations on them. It might seem advantageous always to take the easy route: to learn that new subject from the standard well-tried textbook or from the experienced lecturer. It might be, but beware. Experience shows that it is in that process of first learning the subject that you are most likely to have new ideas about it. Once you allow that process to be entrained by an influential standard approach invented by someone else, you are relinquishing an opportunity to see it in a new way. The globalization of education, which allows many more people to be taught by one person, using video links, has many obvious benefits. But at an advanced level it also has its downside.

Some scientists have invested their careers in some enterprise which hopes for a dramatic change in our scientific fortunes. They may be hoping for a sudden improvement in our technological capabilities in order to allow tiny effects to fall within experimental reach. They may be engaged in the search for extraterrestrial intelligence, a search whose expectations are predicated upon certain optimistic assumptions about the long-term history of civilizations. Progress is something they rely upon as a universal factor in the Universe. Without it they would have to pack up their bags and go home.

Commentators who are not active scientists often have strong views on the future of science. They are no less immune to psychological links. Nonscientists often like to feel that there are, and ever will be, things beyond science. This may be a religious impulse and be developed in quite a sophisticated form, like the 'God-of-the-gaps' argument for the existence of a God. There is also sometimes a jealousy of any activity that seems to be a little too successful. Nor is this all-too-human response confined to commentators. Scientists who have invested years of time and vast amounts of energy in a line of inquiry that demonstrably fails rarely change direction and pursue a new one. They find it difficult to accept the new scheme (they often see it as less imaginative in conception than the unsuccessful one that they must relinquish) and they tend to become hypercritical about it, demanding standards of proof that they would never have required of their own conception.

Summary

How could a mechanism composed of some ten billion unreliable components function
reliably while computers with ten thousand components regularly failed.

JOHN VON NEUMANN[33]

In this chapter we have discussed some of the ways in which we might encounter limits, imposed by our humanity, to what we can know of the Universe. Our minds were not designed with science in mind, nor did evolution primarily fit them for that purpose. We possess the physical and mental attributes that we do as a result of an erratic process of adaptation to ancient environments whose challenges do not confront us today. We are a package of abilities for social interactions, finding safe habitats, finding food, avoiding getting too hot or too cold, attracting mates, keeping out of the way of hazards and predators, and having as many offspring as possible. We have to understand our scientific reasoning ability as a by-product of abilities selected for other, seemingly much more mundane, purposes. Thus, on the face of it, there is no reason why we should possess the conceptual ability to make sense of the way the Universe works. It would require a coincidence of cosmic proportions if the Universe were complicated enough to give rise to life, yet simple enough for one species to understand its deepest structure after just a few hundred years of serious scientific investigation. There is no reason to expect the Universe to have been constructed for our convenience.

We saw how this pessimistic forecast might be avoided. It might be that the exotic concepts that seem to be required in order to understand the Universe can be constructed, bit by bit, from very simple basic ideas like counting, cause and effect, either–or, which our minds do seem to have inherited because of the advantages that an appreciation of such concepts seems to offer in the survival game. When we consider the capabilities of machine computation, we see how much can be achieved by the repetition of very simple instructions.

While this might help us close any gap in sophistication that might exist between our own mental capacity for conceptualization and that required to understand the Universe, it is not the end of our problems. We are faced with unravelling enormous complexities in the states that Nature has created in the Universe by using very simple laws.

The business of understanding the Universe requires us to enlist the help of computers to simulate the way in which the most complicated and long-winded natural processes work. Unfortunately, this confronts us with huge problems of tractability. Even the problems which can be solved by step-by-step computation include many quite small problems which require huge periods of time (far longer than the age of the Universe!) for their solution. These practical difficulties will place a strong limit on our ability to predict, to reproduce, to explain, and to understand the workings of the Universe unless we can find an entirely new method of simulating the behaviour of natural processes.

We saw that the human scientific enterprise is not a single mind. It is a collective activity that can often overcome the limitations of individual minds by collective, linked activity. This is a strength that we can amplify enormously by artificial means. In the future, the increasing use of networked computers will provide us with a powerful tool to overcome our individual limitations. In effect, we shall be evolving, artificially, a large-scale version of the human brain. We saw how the creation of international computer networks has begun this process. But while there are advantages to this evolutionary development, there are pitfalls. The growth of connectedness can lead to an ironing-out of diversity. And, while there are advantages, they are not a panacea. There remain intractable problems that require so much computational time to solve that they are for all practical purposes insoluble.

CHAPTER 5

Technological limits

Do you believe that the sciences would ever have arisen and become great if there had not beforehand been magicians, alchemists, astrologers and wizards, who thirsted and hungered after abscondite and forbidden powers?

FRIEDRICH NIETZSCHE[1]

Is the Universe economically viable?

For which of you, intending to build a tower, sitteth not down first, and counteth the cost, whether he have sufficient to finish it?

ST. LUKE[2]

When considering our possible futures it is easy to consider progress entirely idealistically: to think that everything that can be done will be done. We know it is not so. Western democracies have become increasingly constrained by economics. There are many great scientific investigations that we would love to embark upon, if only we had unlimited money. These questions are brought to a head occasionally when decisions have to be taken about very costly projects, when medical errors require compensation, or when disaster strikes a project like the *Ariane-5* rocket launch and hundreds of millions of pounds are wasted in a split second.

There are many things which might be achievable in principle but which are unachievable in practice. In this chapter we shall explore some of these practical limitations. Some boil down to problems of financial cost; but not all of them do. 'Cost' has a more general interpretation which allows us to analyse many forms of acquiring knowledge. The second law of thermodynamics tells us that we need to do work to acquire information.[3] This way of thinking is very general and allows us to quantify the cost of any computation. In any sphere of human activity it is not enough to be in possession of a procedure to solve a problem. You need also to know the cost of its implementation, either in terms of time, money, energy, or computational power. This knowledge opens up the possibility of finding a procedure that is better, in the sense of being more cost-effective.

A consideration of the cost of acquiring knowledge figures very little in pre-twentieth-century discussions of the progress of science. Today, it is a dominant consideration. In the past, experimentation was fairly inexpensive and on a small scale. Gentleman scientists built their own instruments, and experiments were on a scale that would be commonplace in high-school science classes today. The industrial revolution began the merger of science and technology, leading to the investigations of subatomic physics which have characterized twentieth-century physics, and sparked the growth of large experimental teams and 'big science'. It is primarily in such an environment that cost becomes a key feature of the viability of research projects. Four hundred years ago the problem of exploration would have been similar. In principle, one could send fleets of ships to explore the seas without limit, but in practice these voyages needed sponsors, and sponsors needed some returns on their investments.

These considerations of cost and utility dominate the technological view of science. In the nineteenth century the concept of progress was a pivotal one. The rapid industrialization of the European nations was a consequence of technological developments. The discovery of the principle of evolution by natural selection by Darwin and Wallace[4] complemented this by providing another concept of progress. Inspired by these ideas, philosophers like Nietzsche and Spencer[5] propounded a view, sometimes called the 'Faustian' view of science, which saw humanity as involved in a constant struggle to subdue the forces of Nature and harness them for profitable use. Nietzsche believed that this derived from a deep-seated inherited instinct to control the environment which had been one of the factors leading to human survival in the dim and distant past. Human beings were distinctive in their desire and ability to transform their environments to advantage. This 'will to power', as he called it, was now to be seen in the way technology was employed to control the nature of the environment by technological means.

This is a extreme view. While there is no denying that it has always been one of the motivations for science, it is not the only one. Yet, this manipulative perspective provides an illuminating way of looking at the scope of our applied scientific activities and their spin-offs. In this chapter we shall look at the progress we have made in manipulating Nature, before considering some of the limitations that might curtail our activities. Before we see how far our manipulations have reached, we should look more closely at our position in the cosmic scheme of things. By appreciating our size and location in cosmic history we can understand, more clearly, why and how we need to manipulate Nature artificially if we are to understand its structure.

Why we are where we are

Life is not a spectacle or a feast; it is a predicament.

GEORGE SANTAYANA

Biochemists believe that the level of complexity required to qualify for the title 'life' can evolve spontaneously only if it is based upon the unusual chemical properties of the element carbon.[6] This is not to say that living complexity cannot *exist* with another element at its base, only that non-carbon life would require carbon-based life to seed or initiate it. For example, at the moment we can see that a rather particular form of organized complexity, based upon the physics of the element silicon (rather than its chemistry), is moving towards becoming organized in such a way that it qualifies for the title of 'artificial life' or even 'artificial intelligence'. But this development has occurred only because of the assistance rendered by a pre-existing form of carbon-based life—us!

Carbon is made in the stars. The simple elements of hydrogen and helium, which make up 99.99999 per cent of the matter in Universe, originated in the first few minutes of the expansion of the Universe, when it was vastly hotter and denser than it is today.[7] One of the great successes of the remarkable 'Big Bang' theory of the present structure and past history of the Universe has been its ability to predict successfully the abundances of these and other light elements that were produced by nuclear reactions in its early stages. The hydrogen and helium from the early Universe is burnt within the interiors of stars into heavier, biological, elements like carbon, oxygen, nitrogen, and phosphorus. When the stars reach the ends of their lives they explode and disperse these building blocks of life throughout space. There, these elements condense into grains and planets. Ultimately, they find their way into our bodies.

This process, whereby the primordial elements from the Big Bang are transformed into the possible building blocks of biology is a long slow business. It takes billions of years. This simple fact reveals something mysterious and important about the Universe around us. It shows us why it is so big.

The Universe is expanding; the distant clusters of galaxies are receding from one another at a rate that increases with their distance apart. This means that the questions of the age and size of the Universe are inextricably linked. The fact that the Universe must be billions of years old, in order to have had enough time to produce the biological elements which make the spontaneous evolution of complexity possible, means that any life-supporting universe must be billions of light years in size. The Universe needs to be billions of light years in size in order to support just a single outpost of life.

Other striking features of the observable universe follow from this simple argument. The vast age and size of a life-supporting Universe make the darkness of the night sky and the coldness of space inevitable features of such a world.

The expansion necessarily lowers the density of matter and radiation in the Universe to very low levels. There is not enough energy to make the night sky bright, and galaxies and stars end up being separated by vast distances. Ironically, the fact that the Universe appears big and old, dark, cold, and lonely is a feature that is necessary for it to provide the building blocks of any form of chemical complexity.

If we look more closely at our environment we can understand its general characteristics. Suppose we were to catalogue all the most impressive natural aggregates of matter that we have encountered in the Universe, from the world of subatomic particles right up to the realm of galaxies. If we locate them on a graph according to their mass and average size the result is as in Fig. 5.1.

Remarkably, all the objects shown in Fig. 5.1 lie along a band running from bottom left to top right. The rest of the picture is blank. There is no mystery about this. The band is the line of constant density.[8] For the structures shown, it corresponds to atomic density: the density of objects which are made of collections of atoms. Their density is very similar to that of a single atom. Only at the top right of the diagram, beyond the scale of stars, do the structures start to fall short of this line. This is because clusters of stars and galaxies are not solid atomic bodies but collections of bodies orbiting under conditions where there is a balance between their mutual gravitational attractions and energies of motion.

The location of the objects along the constant density line is no accident. They mark out the places where different types of balance are possible between opposing forces. Take planets for example. A habitable planet needs to be large enough for the strength of its gravity to retain an atmosphere, yet not so great that the force of gravity at its surface will break the delicate chemical bonds that hold complex biochemical molecules together. These opposing considerations narrow the range of sizes that a habitable planet can have, and render many of its surface properties inevitable. If it is necessary for a significant region of a habitable planet's surface to exist for long periods at a temperature where water is liquid, then the range of planetary properties which allows life to arise may be smaller still. In fact, more complicated aspects of the planet's motion around its parent star are equally important in this respect. The orbit of the planet must be a suitable distance from the star to maintain temperate life-supporting conditions. This requires the average distance of the planet from the star to lie within narrow bounds, and the shape of the orbit must not deviate too greatly from tracing out a circle. Furthermore, as the planet rotates, the tilt of its axis of rotation with respect to the plane in which it is orbiting must not be too great or the seasonal climatic changes will be oppressively large. Huge changes in sea level and glaciation would create an environment in which the part of the planetary surface most conducive for the evolution of advanced life would be very small.

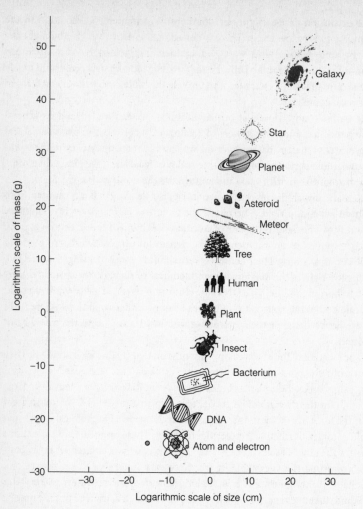

Fig. 5.1 The pattern of masses and sizes for the most distinctive structures in the Universe.

Some consequences of size

> *One should not exaggerate the importance of trifles. Life for instance, is much too
> short to be taken seriously.*

NICOLAS BENTLEY

The strength of gravity on a habitable planet's surface determines how large
living things can become before their size becomes a liability. Strength does not
increase at the same rate as weight and volume of a structure grows. A horse

cannot carry another horse on its back, but a small dog can easily carry two similar dogs on its back, and an ant can carry a load many times greater than its own body weight. (For an interesting demonstration of this try the variation of the world weight-lifting records versus the weight of the lifter: see Fig. 5.2). If you try to scale up organisms in size, you will eventually find that they are too weak to support their own weight. They would just collapse under the pressure.

Our own size is interesting in a number of ways. It appears to have been increasing gradually over the course of human evolution. On the largest scale, we see that we lie midway between the astronomical and subatomic realms. Locally, we see that while we have an unremarkable position on the size spectrum of the Earth's living creatures, we are distinguished by being the largest of them that walks on two legs. It also appears that our size has been crucial to the pattern of social and technological development that we have followed. Our size makes us strong enough to break molecular bonds in solid materials. We can shatter and etch stones and sharpen hard materials like flint. We can bend and fashion metals. We can throw rocks and wield sticks with sufficient kinetic energy to kill other members of our species as well as other animals. This capability, which we would not obviously possess if we were significantly smaller, has an important role to play in our evolution. It has made early technological development possible, but it has also made us a dangerous warlike species, able to exert deadly force very easily. It made rapid progress

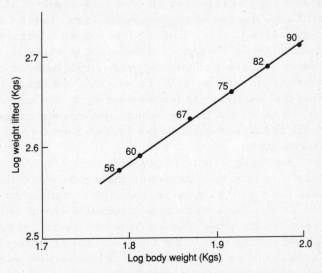

Fig. 5.2 The variation in the world weightlifting records with the weight category of the lifter. There is an excellent fit to the line which denotes the variation of weight lifted (strength) to the two-thirds power of the lifter's weight, which is proportional to his strength.

possible, but it has always provided us with the means to bring all progress to an end.

Another curious consequence of our relatively large size is our ability to use fire for many specific purposes. No other animals can do this. There is a smallest flame defined by the balance between the volume of combustible material available and the surface area over which oxygen can fuel the combustion reaction. As the volume of combustible material gets smaller, the surface becomes too small for the flame to persist, and it dies. Ants could not use fire because they would find the smallest stable flames too large for them to approach safely and feed the fire. Small stable flames are well suited to the needs of creatures of human size. (This does not guarantee that they will make use of them though; chimpanzees make no use of fire.) Its use has all sorts of important consequences. It enables more hours of the day to be used; it offers protection from predators; it improves health and diet, because cooked food harbours fewer dangerous bacteria and is more digestible. The development of cooking also increases the range of palatable foodstuffs and allows more time to be employed on activities other than hunting.

Finally, there is a close correlation between size and lifetime in living creatures. A large animal is a considerable investment of scarce resources and needs to have a long lifetime in order to be worth the investment in evolutionary ('selfish gene') terms. As a result, large animals tend to have small litters and to nurture their young with great care and attention. Small animals adopt a different strategy. They have large litters and short lifetimes. There is a lower probability of survival for any individual, but this is counterbalanced by the larger number of young. The large size of humans is linked to their relatively long lifetimes and their extraordinarily long periods of childhood, during which they are cared for by their parents. This has many social consequences. Long periods of close interaction with family and community members lead to complex social relationships. Learning can be extensive, and groups can acquire considerable amounts of environmental knowledge which they can pass on to their close companions.

We have seen from Fig. 5.1 that our location in the cosmic scheme of things is determined in a general way by the relative strengths of the forces of Nature. Our actual size range is a consequence both of these principles and of a sequence of historical accidents and complex interactions with our environment and with other species.[9] This is important for our study of human technical progress and its limits, because it shows why we need technology. If we are to manipulate the environment on scales much larger or smaller than our own body size, then we must do it by artificial means.

Our size determines our strength, and hence the extent to which we require artificial aids to construct or break bonds created by the forces of Nature. Our environment determines the keenness of our senses and the extent to which we need to supplement them in order to explore the world in greater detail. Only an extraordinary coincidence between our technological reach and the conditions of Nature that we wish to investigate will avoid a technological barrier to human scientific understanding.

The forces of Nature

The machines that are first invented to perform any particular movement are always the most complex, and succeeding artists generally discover that with fewer wheels, with fewer principles of motion than had originally been employed, the same effects may be more easily produced. The first philosophical systems, in the same manner, are always the most complex.

ADAM SMITH[10]

We have long been aware of the forces of gravity and magnetism. No land-dwelling creatures could be unaware of gravity. But magnetism became known to us only because of the presence of the Earth's magnetic field and the discovery of magnetized metallic ores at the Earth's surface. In modern times, the discovery that moving magnets can produce a flow of electrical current and, conversely, that under suitable conditions electrical currents can create magnetism has led us to appreciate that there is another force of Nature at the root of both phenomena. We call it electromagnetism. It behaves in the same way in every place we have observed it in the Universe and our mathematical understanding of it is extraordinarily accurate—good to one part in 10^{11}. To use an analogy suggested by Richard Feynman, one of the creators of this theory, this is equivalent to being able to measure the distance from London to New York with an accuracy better than the width of a human hair.[11] We can predict the behaviour of gravity with even greater accuracy. We can monitor the motions of a pair of neutron stars 30,000 light years away, in an environment where the strength of gravity is 10^5 times stronger than anywhere within our solar system. Over a period of more than 20 years we have found that these movements agree with the predictions of Einstein's theory of gravity to the limiting accuracy of our measurements[12]—good to an astonishing level of one part in 10^{14}.

Gravity and electromagnetism are not the only forces of Nature. So far, we have discovered two others: the 'weak' force and the 'strong' force. The weak force is the source of radioactivity, while the strong force binds the nuclei of the chemical elements together. When this binding energy is released, enormous nuclear energies are available. These forces have very short ranges and their terrestrial effects can be isolated from those of gravity and electromagnetism only in special circumstances which require considerable technological expertise to engineer. However, in the astronomical realm, these forces play a more dramatic role. They are responsible for energy production in the stars, for the stability of the Sun, and hence for the existence of any form of planet-based life like ourselves. In order to probe the basic characteristics of the strong force we have to study distances of 10^{-13} centimetres; to probe the weak force we have to go a hundred times smaller, down to 10^{-15} centimetres.

At present, these four forces of Nature are the only ones known. Remarkably, we need only these four basic forces to explain every physical interaction and structure that we can see or create in the Universe.[13] Physicists believe that these forces are not as distinct as many of their familiar manifestations would seduce us to believe. Rather, they will be found to manifest different aspects of a single force of Nature. At first, this possibility seems unlikely because the four forces have very different strengths. But in the 1970s it was discovered that the effective strengths of these forces can change with the temperature of the ambient environment in which they act. The weak and the electromagnetic forces change in strength as the temperature rises, and they become similar in strength when a temperature of about 10^{14} degrees Kelvin is reached. They then combine to generate a single 'electroweak' force with two aspects. By contrast, the strong force is observed to weaken, and it starts to approach the electroweak force as the temperature increases further. Eventually, they should all become of equal strength, but only at a huge energy of about 10^{15} giga electronvolts (10^{28} degrees Kelvin) that lies far beyond the reach of terrestrial particle colliders (Fig. 5.3).

The fourth, and most familiar, force is gravity. Yet, despite its ubiquity, gravity is deeply mysterious in structure and peculiar in its relation to the other three forces. It is closely entwined with the character of space and time. Altering the strength of gravity can alter their properties. At present, we have an attractive mathematical theory ('superstring theory') which provides a way of under-

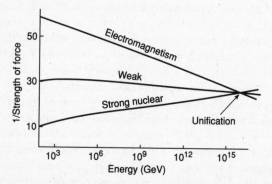

Fig. 5.3 The predicted variation of the strengths of the strong, electromagnetic, and weak forces of Nature in a supersymmetric theory of their interactions. A suggestive crossover ('grand unification') is predicted at high energy.

standing how to unify gravity with the other three forces of Nature. This possibility comes about only at the highest temperature that can exist in Nature, and may have ever existed only when the Universe came into being. As we shall see, this creates problems for the testing of any ultimate 'Theory of Everything'.

The foregoing short description of our position in the Universe relative to the scale of astronomical structures and the most elementary particles of matter shows what we can and cannot expect to learn about the world. There are simple physical principles which constrain the sizes of living things on planetary surfaces. Our size means that we can easily learn about the superficial properties of things on the scale of centimetres or metres. But if we are to understand the structure of astronomical bodies and the realm of molecules, atoms, and smaller phenomena, then we need to use artificial aids. If we are to discover the full story about the forces which govern the subatomic realm of elementary particles, then we must catch Nature in circumstances far removed from the limits of our senses.

The difficulties that we shall encounter in furthering our control over Nature are therefore a consequence of our size and the properties of an environment able to support planet-based atomic life forms. Living beings must find themselves inhabiting environments which are friendly for organized complexity. This means environments that are cool enough to leave molecular bonds intact, yet warm enough for water and other simple liquids to exist. We might expect any other intelligent planet-based life-forms to have encountered similar limitations, and to have overcome them in a manner not dissimilar to our own.

An interesting by-product of a life-supporting environment is the way in which the presence of an atmosphere creates colours by scattering light from the parent star (in our case the Sun). The presence of colour drives natural selection to favour adaptations which can exploit its detection to advantage. The growth of a pigmented flora produces colour variations in sources of food and endows adaptive advantage to simple forms of colour discrimination. Any rotation of the planet, which is hard to avoid, creates daily variations in light levels. In this way the size of the planet and its atmosphere, together with particular molecular properties, are the sources of the adaptive evolution of colour vision. Regardless of the final form of that visual sense, it is clear that its capabilities will be an adaptation to the local environment, not to the requirements of studying the larger and smaller properties of the Universe that science requires. We are thus endowed with senses which had vital roles to play. They needed to sense danger, distinguish edible fruits from green foliage, and detect the daily changes of light and twilight. Any species coevolving in that environment by natural selection will have limited sensual ranges, and will need to manipulate Nature by artificial means if it is to escape from the prison imposed by its senses.[14]

For some, that prison may be darker than for others. We are fortunate to have clear skies for most of the time. This makes astronomy possible. From its study has flowed our understanding of gravity and many other aspects of terrestrial physics which become noticeable only when viewed in the astronomical realm. The element helium, which plays such a key role in our investigation of the behaviour of matter at low temperatures, was first discovered in the coronal spectrum of the Sun during an eclipse by the French astronomer Pierre Janssen in 1868 (hence its name).[15] If our skies had been cloudy and overcast, then naked-eye and optical astronomy would not have developed. We might have little or no knowledge of our own solar system, let alone the myriads of distant galaxies beyond. Life could evolve to a level similar to that displayed by *Homo sapiens* on a planet that failed to provide any metallic ores at its surface. But the subsequent development of technology would be stalled and a long, albeit sophisticated, stone age might be a feature of that planet's evolution. Quirks of geology provided us with the means to develop technologies capable of probing the microscopic structure of matter and discovering electricity.

The simple lesson we learn from this is that different intelligences in different parts of the Universe will have evolved to meet the challenges posed by their specific environments. Those environments will determine the directions of developments which are most reproductively advantageous and least expensive in terms of life's scarce resources. They will also define what is impossible for its inhabitants. This, in turn, will influence the directions in which they strive for knowledge and progress.

Manipulating the Universe

> *Philosophers can be divided into two classes: those who believe that philosophers can be divided into two classes, and those who do not.*
>
> ANONYMOUS

The Faustian picture of humanity engaged in a quest to manipulate Nature over a wider and wider domain is an incomplete one, but it is a useful way of organizing our picture of human technological achievement. We can gain some perspective on where we have got to, and how far we have to go, by looking at how well we have done in manipulating matter on scales larger and smaller than those of our own bodies, in the two realms where we need assistance by artificial aids.

In the 1960s the idea of searching for extraterrestrial intelligences (ETIs) was a new and novel one, with many new techniques of astronomical observation potentially at its disposal. A Russian astrophysicist, Nicolai Kardeshev, proposed that we divide advanced ETIs into three categories of Type I, Type II, or Type III, according to their technological abilities.[16] These grades of civilization were loosely distinguished as follows:

Type I is capable of restructuring planets and altering its planetary environment; it can use the present energy equivalent of terrestrial civilization for communication;

Type II is capable of restructuring solar systems; it can use the present energy equivalent of the Sun for interstellar communication;

Type III is capable of restructuring galaxies. It can signal across the entire observable Universe using laws we know; it can use the present energy equivalent of the Milky Way Galaxy for interstellar communication.

The motivation for this classification was to estimate how much waste heat might be produced by the technological activities of these civilizations so that one could decide if any could be detected by astronomers.[17] This reveals whether it is easier to detect a very distant Type III civilization than a nearby Type I.[18] But we are not primarily interested in this aspect of Kardashev's classification. Rather, we want to extend it so as to define a ladder of technological milestones of achievement.

In this scheme of things we can see that we are certainly a Type I civilization. We have altered the topography of the Earth's surface in many ways: building structures, mining and excavating, removing rainforests, and reclaiming land from the sea. Our industrial activities have altered the behaviour of the Earth's atmosphere and changed the temperature of the Earth. We have the capability to make major changes to the Earth and its immediate environment, either by design or by accident. Our exploration and exploitation of the Earth's interior structure has been relatively modest so far, and amounts to little more than the extraction of fossil fuels and minerals.

We are nearly a low-level Type II civilization. We could alter the evolution of some of the inner planets; (for example by seeding Venus with primitive life forms, which would alter the atmospheric chemistry) and we could (indeed, we may have to) apply a form of Star Wars technology to protect ourselves from incoming asteroids and comets when they are in the outer solar system. A mature Type II civilization might be engaged in altering the chemical composition of their local star in some way (perhaps by diverting comets into it) in order to change the nature of their own biosphere. Such a civilization could harvest minerals and heavy elements from space and learn to extract solar energy with much higher efficiency than our present technology allows.

Type III civilizations are the stuff of science fiction stories, and it is hard for us to conceive of manipulating matter over such enormous dimensions (perhaps by affecting the operation of cosmic radio jets—the largest coherent structures seen in the Universe) because of the long periods of time that are necessary for signals to traverse these dimensions.[19] In order for a civilization to find such fantastic foresight advantageous, it would have to have all possible local problems completely under control, and have very long (even unending) individual lifetimes. If local environmental problems still presented significant challenges, then it is unlikely that profligate scientific adventures would be embarked upon. However, when those challenges had grown into threats to the whole civilization's future existence, all resources might be thrown into a search for a means of relocating to a safer environment.

At first, it might appear that costly, ultra-long-term projects would never be embarked upon if they greatly exceeded the average lifespan of an individual member of society. How might this disincentive be overcome? One could imagine that the concept of an individual lifetime might become irrelevant. With very sophisticated computer technology, capable of making complete 'back-up' copies of minds, individuals could overcome 'death' in the usual sense. They might miss a brief interval of time while information was transferred to a new medium, but this would be only a minor diversion. One could imagine different computers vying to provide the fullest regeneration, the one that lost the least experience, as opposed to the ones that removed some unwanted attributes, or 'bad' memories, at the same time.

In recent years there have been detailed speculations about the far future of the Universe that envisage the existence of beings even more advanced than those of Type III.[20] Suppose that we extend the classification upwards. Members of these hypothetical civilizations of types IV, V, VI,..., and so on, would be able to manipulate the structures in the Universe on larger and larger scales, encompassing groups of galaxies, clusters, and superclusters of galaxies. Ultimately, we could imagine a type Ω civilization, which could manipulate the entire Universe (and even other universes). If time travel is a practical possibility, then its achievement would open up a whole new world of possibilities for civilizations of this ultimate type. They would be defined by their ability to reach as close as possible to all the fundamental limits on information storage, processing, resistance to chaotic unpredictability, and endurance.

There has been much detailed speculation about what a Type Ω civilization might in principle be able to do, and how it might do it. Together with Frank Tipler, the author has shown that it is possible for information processing to continue indefinitely into the future in certain types of expanding universe, and has also shown that there need be no barrier to the extent of its influence if the Universe possesses a certain type of overall structure.[21] However, these studies merely identify the best possible situation; it is quite another matter to achieve it.

In the same vein, Alan Guth has explored what might need to be done in order to create a 'universe' in the laboratory.[22] This does not seem to be possible as we understand physics at present, but relatively small changes to our knowledge might make it a technical possibility in the far future. One should add that we would not be able to see or interact with this 'baby universe' after it was initiated. In the most general versions of the inflationary universe scenario, which we shall explore in the next chapter, it is even possible for different regions of the Universe to be endowed with different values for the local constants and laws of Nature.[23]

Lee Smolin has considered a speculative scenario in which the values of the constants of Nature evolve through many 'editions' as new universes emerge from the collapse of black holes, with small shifts in the values of their defining constants occurring at each stage.[24] The shifts are 'selected' by the propensity to increase the production of black holes, and so create more opportunities for subsequent baby universes with slightly shifted constants. The motivation for this proposal was to provide an explanation for the many peculiar coincidences that are observed to exist between the values of different constants of Nature. These coincidences are the outcomes of a process that has continued for an enormous period of time. As the outcome of a selection process, the finely balanced situation that we observe could be imagined to be optimal with respect to the selection process. In this case, any change in the values of the constants away from their observed values should lower the black-hole production rate.[25]

With the whole spectrum of possibilities like these in mind, the American cosmologist Edward Harrison has raised interesting questions about the extent to which intelligent beings could influence the values of the constants of Nature that define the character of their Universe.[26] Their civilizations would be of the Type Ω variety. Harrison speculates that the fact that so many of the constants of Nature take values which seem remarkably suitable for the evolution of life might be a consequence of the ability of successive generations of advanced civilizations to create expanding 'universes' and engineer the values of their physical constants to approach the optimal values for the subsequent existence and persistence of life. Let us consider his idea more closely.

We know that the constants of Nature which define our own Universe are tantalizing in many respects.[27] We do not have an explanation for the values that they possess, but, were some of those values slightly changed, it would not be possible for organized complexity (of which life is an extreme example) to exist. There are a number of famous and finely balanced coincidences relating to the values of the natural constants; and if these coincidences did not exist then neither could we. As yet, we do not know whether these coincidences are just lucky outcomes of all (or a wide range of) possibilities, or whether there is one and only one possible combination of values for the constants of Nature that is logically self-consistent. If other values of the constants are possible, and early investigations of candidate superstring 'Theories of Everything' imply that they are, then the values of constants might be tuneable if universes could be 'created' experimentally from vacuum fluctuations. Any civilization that was technologically advanced enough to do this might tune the constants to be a little more conducive to the evolution of life than they found them to be in their own universe. After many generations of tuning by successive advanced civilizations, we might expect the constants to possess finely tuned values that were close to optimal with respect to the conditions that are needed to allow life to arise and evolve successfully. The fact that our own Universe possesses what some regard as a suspiciously good fine tuning might even be regarded as evidence that this successive tuning of long-lived universes by advanced inhabitants has been going on for many cosmic histories already. Unfortunately, this amusing idea cannot explain why the constants were such as to allow life to originate long before the ability to tune baby universes existed. It requires us to believe that life was fortunate to find the universe so hospitable, or that life is virtually inevitable, for a huge range of values of the constants of Nature, in which case it is hard to understand why the Type Ω civilization would go to great lengths to tune the constants. But maybe great lengths are unneeded.

The British cosmologist Fred Hoyle once responded to his discovery of the remarkably fortuitous location of energy levels in the carbon and oxygen nuclei, without which our existence might well be impossible, by offering the following bold opinion:

> I do not believe that any scientist who examined the evidence would fail to draw the inference that the laws of nuclear physics have been deliberately designed with regard to the consequences they produce inside stars. If this is so, then any apparently random quirks have become part of a deep-laid scheme. If not then we are back again at a monstrous sequence of accidents.[28]

A similar teleological suspicion is found in Freeman Dyson's reaction to further coincidences about the strengths of the electromagnetic and nuclear forces,[29] which prevent nuclear reactions consuming the material of the stars so rapidly that life-supporting environments disappear long before evolution can produce biological complexity:

> As we look out into the Universe and identify the many accidents of physics and astronomy that have worked together to our benefit, it almost seems as if the Universe must in some sense have known that we were coming.[30]

We should stress that the ideas of Smolin and Harrison are extremely speculative, but they provide examples from our own limited imaginations of some ways in which a Type Ω civilization might go about influencing the fabric of the Universe in the far future.[31]

We have introduced a classification of civilization 'types' by considering their ability to manipulate the *large-scale* world around them. This is the hardest manipulation to conduct. It requires huge energy resources and is very difficult to reverse if things go wrong. Gravity is inevitably involved, and because it is the only known force of Nature that acts on everything, without exception, it cannot be switched off. In practice, therefore, we have found it much more cost-effective to extend our ability to manipulate the world over smaller and smaller dimensions rather than over larger and larger ones. So, let us extend our classification of technological civilizations downwards as Type I-minus, Type II-minus, . . ., and so on, down to Type Ω-minus, according to their ability to control *smaller and smaller* entities. These civilizations might be distinguished as follows:

Type **I-minus** is capable of manipulating objects over the scale of themselves: building structures, mining, joining and breaking solids;

Type **II-minus** is capable of manipulating genes and altering the development of living things, transplanting or replacing parts of themselves, reading and engineering their genetic code;

Type **III-minus** is capable of manipulating molecules and molecular bonds, creating new materials;

Type **IV-minus** is capable of manipulating individual atoms, creating nanotechnologies on the atomic scale and creating complex forms of artificial life;

Type **V-minus** is capable of manipulating the atomic nucleus and engineering the nucleons that compose it;

Type **VI-minus** is capable of manipulating the most elementary particles of matter (quarks and leptons) to create organized complexity among populations of elementary particles;

culminating in.

Type **Ω-minus** is capable of manipulating the basic structure of space and time.

Again, we can attempt to locate ourselves in this classification of technical capability. We have long been a Type I-minus civilization, and modern genetics allows us to be a Type II-minus in several respects. The use of this ability is controversial and fraught with dangers and possible abuses of civil and personal liberties. The Human Genome Project is an international research programme to decode human genetic information with a view to identifying causes of various human traits and medical disorders. It marks the entry of biology into the multinational 'Big Science' league previously dominated by physics and astronomy.

We also have some Type III-minus abilities, and routinely engineer materials to possess particular structural features; medical scientists design antibiotics to have special therapeutic properties. We have only just entered the Type IV-minus domain. Although we are beginners, we have developed an ability to move individual atoms and engineer surfaces at the level of single atoms (see Fig. 5.4).

This ability forms the basis of the quest to develop nanotechnologies. It has long been a dream of scientists to construct microscopic machines—motors, valves, sensors, and computers—down at the molecular scale. They could be implanted into larger structures where they would carry out their invisible function, perhaps monitoring the heart of a cardiac patient or keeping vital arteries clear of blockages. Some devices of this sort already exist (see Fig. 5.5). They are likely to play an increasing, but unseen, role in everyday life in coming years.

We are struggling to maintain our status as a Type V-minus civilization. We have been able to utilize nuclear forces and particles in controlled ways to create sustained energy by nuclear fission, detonated explosions by nuclear fission and fusion, but have failed to control all the by-products of these actions safely and reliably. Despite long and expensive investigations in many countries, we have

Fig. 5.4 The atomic corral. Forty-eight iron atoms forming an enclosure of radius 0.01 microns on a copper surface. The atoms were positioned using a scanning tunnelling microsocope.[68]

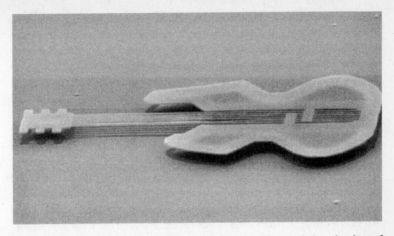

Fig. 5.5 A guitar, twenty times smaller than the width of a single human hair, sculpted out of crystalline silicon using an electron beam. It has six strings, each one hundred atoms thick which can be plucked with an atomic force microscope. The frequencies produced are too high for the human ear to detect. (Photo by D. Carr and H. Craighead, Cornell University, New York.)

failed to produce viable sources of controlled energy from nuclear fusion reactions. Although this is a safer and cleaner source of nuclear power than nuclear fission, it presents formidable problems of confining and controlling the plasma of interactants. So far, the controlled power output has lasted for only very brief periods and the process is far more expensive than conventional energy sources. It is, however, likely that these problems will one day be solved. In fact, the Italian physicist Carlo Rubbia has outlined ways in which clean sources of energy could be obtained by the high-energy bombardment of nuclei with fast particles. As a bonus this technique offers a simple means of rendering radioactive nuclear waste harmless, and (unlike existing reactors and processes) produces no by-products of any military use.

Another recent success of the Type V-minus sort has been the creation of a nucleus of antimatter (antihydrogen) at CERN in Geneva. Ultimately, the controlled meeting of matter and antimatter could provide us with clean, safe energy. The challenge, as usual, is not merely to do this, but to do it economically enough to make it worth while.

We are not yet a Type VI-minus civilization. We can produce elementary particles in high-energy collisions between protons and in other high-energy particle physics processes, but we are still at the stage of watching the debris from those events in order to advance and consolidate our knowledge of the elementary particles themselves: to understand how many of them there are, their masses and lifetimes, and the qualities that identify them, and limit the scope of their mutual interactions. As yet, we are unable to engineer these particles to produce complex aggregates with particular properties (femto engineering?). We do not know whether such structures can exist in forms other than the known aggregates which make up hadrons and mesons.

It is worth remarking that these manipulations of the smallest components of matter give rise to a remarkable state of affairs. We have precise mathematical theories which predict the behaviour of the microscopic world with unprecedented precision. These theories predict more things about the world than we have already learnt by observing it. Occasionally, they allow us to engineer a very peculiar situation that may well never have arisen anywhere in the Universe during its entire past history, unless other sentient beings have conducted similar experiments. For example, in the early years of this century the phenomenon of superconductivity was first observed in Leiden, in 1911, by the Dutch physicist Heike Kamerlingh-Onnes. He observed the disappearance of all resistance to the flow of electrical current in mercury cooled to -269 degrees Celsius, just 4 degrees above the absolute zero of temperature (-273 Celsius). There is no reason to expect such extraordinarily low temperatures to exist naturally anywhere in the Universe. If not, then the phenomenon of super-conductivity may never have been manifested in the Universe before its appearance in Leiden in 1911. Likewise, the discovery of high-temperature superconductivity in Zurich in 1987, by Georg Bednorz and Alex Mueller, may have been a first for the Universe. This phenomenon (which is still not understood fully by physicists) occurs at higher temperatures than traditional superconductivity (hence its name), does not have the same physical explanation, and arises in materials that are an unusual cocktail of minerals. The specific chemical mixtures are rather delicate (shades of the alchemists' secret formulae) and there is no reason to expect them to occur spontaneously in natural environments, like a planetary surface, or to evolve from interstellar material. In that case, this phenomenon will have made its debut in the Universe as a consequence of the manipulations of matter under artificial conditions by human beings. This is a very sobering thought.

The ultimate technological challenge for a Type Ω-minus civilization would be the manipulation of space and time.[32] Perhaps they would be able to tap into the quantum zero-point energy of the Universe and use it as an energy source. At present, we can appreciate (theoretically) some of the things that such a super-civilization might be able to do to space and time, but the conditions needed to implement such changes are far beyond the reach of our technology.

Einstein taught us that moving clocks go slow and that clocks go slow in strong gravitational fields. We can observe these things occurring in high-energy physics experiments, showers of cosmic rays from space, and in observations of motions in the solar system and beyond. However, we are not yet in a position to create the circumstances in which these effects would be of technological benefit. A classic example, familiar to readers of science fiction stories, is the possibility of travelling in a short period of the traveller's time to star systems many light years away when reckoned by us back home, by moving at a speed close to that of light. We also appreciate that there might be peculiar configurations of mass and energy which permit time travel to occur, or for local 'wormhole' connections to be forged between parts of the Universe which appear (in terms of conventional light travel times) to be enormously distant.[33]

Possibilities of this speculative sort have features which prevent us dismissing them out of hand. We have a theory of gravitation, the general theory of relativity, which works with fantastic accuracy in every arena where it has been tested. We also appreciate some of its limitations; that is, we know that it must fail under very extreme circumstances of temperature or material density (which we are in no danger of encountering or creating at present). This theory permits things like time travel to occur. But we do not know the full collection of restrictions that we have to impose upon the predictions of this theory in order to pick out those which are compatible with all the other properties of our Universe. Even when we have done that, we have to ask about the likelihood of something occurring. Time travel may be possible in principle, and involve no violation of the laws of Nature, yet have too low a probability of occurrence (because of the very special circumstances required) for it ever to be witnessed in practice. For example, levitation is compatible with the known laws of physics, in the sense that if all the molecules in my body just happen to drift upwards at the same moment, then I will leave the ground. No law of physics forbids this. There is a chance that this freak situation will occur; but that chance is so low that we can be sure that any report of it happening is much more likely to be mistaken than it is to be true.

It is curious that macroscopic and microscopic capabilities of Ω and Ω-minus type civilizations come full circle and join. The ability to control whole universes, or to create them from the quantum vacuum, involves the *microscopic* control of space and time. Universes actually contain nothing other than space and time. All matter can be viewed simply as undulations in what would otherwise be a perfectly smooth substratum of space and time.

Before leaving our classification of technological achievement we should consider whether there is not a third principal category of manipulative achievement. Besides the realm of the very large and the very small, there is the realm of the *complex*. In our experience, the most complex things inhabit sizes intermediate between the very large and the very small. In Fig. 5.6, we have catalogued some of the members of this complexity club. They are distinguished by their internal organization: by the number of interconnections that exist between their sub-components. As the number of those connections

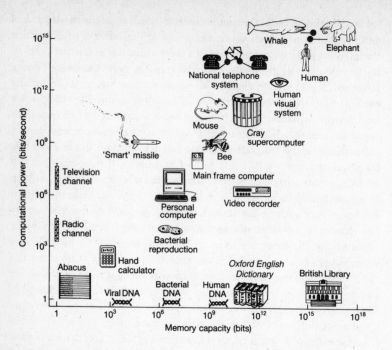

Fig. 5.6 A representation of two aspects of structures, their information storage capacity (in bits), and their processing power (in bits per second). The most complex known structures, which combine large information storage with rapid processing, lie in the top right-hand corner.

increases, so the potential for complex behaviour rises dramatically, in sudden leaps and bounds.

As we delve into the realm of the complex, we find a world that orders itself in ways that are quite different from those of the simple realms of the large and the small. Stability is not just the result of two counter-forces of Nature coming into balance. It is possible for equilibrium to be dynamic, yet steady; for incessant local change to underwrite overall order, far from equilibrium. It is the realm where we find the stability of the candle flame, rather than that of candle wax.

Criticality: the riddle of the sands

> To see the World in a grain of sand,
> And a Heaven in a wild flower,
> Hold infinity in the palm of your hand,
> And Eternity in an hour.

WILLIAM BLAKE[34]

So far, there is no general theory of complexity. Like life, it is hard to define, but we know it when we see it. We witness many particular examples.[35] They share certain common features, but no simple set of laws has emerged which captures the essence of all forms of complexity. This might be too much to hope for; a more realistic possibility might be the discovery that there can exist only a finite number of different varieties of complexity and every example that we discover will fall into one or other of these classes. In recent years, an important form of complex arrangement has been identified which may characterize one of these classes. It displays a type of behaviour that has become known as *self-organizing criticality* (SOC).[36]

The central paradigm of SOC is the simple example of the sand pile. Imagine that grains of sand are dropped one by one on to a flat surface, like a table top, with open edges so that any excess sand can fall over the edges of the table. At first, the sand pile gets steadily steeper; incoming grains of sand affect only the behaviour of other grains in the immediate vicinity of where they fall, and there are only occasional, small avalanches of sand. As the sand continues to fall, the pile does not continue to steepen, though; gradually, the slope approaches a 'critical' angle. This slope is maintained by avalanches of sand cascading down the sides of the pile (see Fig. 5.7).

This critical state has several fascinating properties. It is a complex organized state that has been created by a concatenation of events (falling, tumbling, sand grains) which, individually, are chaotically unpredictable. But this is not the

Fig. 5.7 A sand pile in a critical state.[69]

only surprising feature of the sand pile. It is in a steady state (the rate of sand falling on the pile will ultimately be balanced by the rate at which sand falls off the edge of the table) which is always on the verge of instability; each sand grain that falls produces avalanches and changes which serve to maintain the overall slope. We call such an equilibrium *critical*. It combines a curious mixture of predictability and unpredictability. Although the overall slope of the pile arises irrespective of how the sand is poured,[27] the sand pile becomes increasingly sensitive to the arrival of each grain as it gets closer to the critical state. Thus, at first the incoming grains affect only neighbouring grains, but as the pile approaches criticality the effects of each arriving grain will extend further and further over the surface of the pile. At criticality, grains will often produce avalanches which encompass the whole scale of the sand pile surface.

The sand pile possesses characteristics of many complex organized systems in Nature. The necessary ingredient for a system to self-organize into this type of critical state seems to be that there is a very wide range of behaviour (avalanches of sand) for which there is no special, preferred size.[28] In the sand pile, this means that between the size of a single sand grain up to the size of the whole pile there must not be special sizes of avalanche that form. At the critical state, there are avalanches of all possible sizes, occurring with varying probabilities.[39]

The sand pile seems to be representative of systems which at first appear totally different. If, instead of avalanches of sand, we have extinctions in a complex ecosystem, then the critical state might represent a dynamic state of ecological balance.[40] The extinctions play a positive role, like the avalanches of sand, because they make ecological niches free for new species to take over. Or, we might consider the overall pressure equilibrium at the Earth's surface maintained by volcanoes and earthquakes[41] as an example of SOC. Another interesting example that has come to be studied in detail is that of traffic flow. The optimal state for a congested road system seems to be a self-organized critical state in which traffic jams of all lengths can arise to maintain the optimal traffic flow. Small jams occur inside large jams and a small movement of a single car can have a large knock-on effect.[42] These fluctuations may be irritating when you are a motorist, but they are the means by which the most efficient overall flow of cars is maintained. If you have a flow with fewer fluctuations, then either the road is being underused, with a very low flow of cars, or it is jammed solid with everyone stationary. Thus, at the optimal critical state, we experience the irritating jams that seem to arise incessantly for no reason at all. Indeed, they cannot be traced to any single cause. This is SOC at work, and it is part of the intrinsic unpredictability of events near the critical state. Similarly, if earthquake activity is truly an example of this form of organized critical behaviour, there is no point in trying to predict the occurrence of earthquakes.

There may be other important examples of self-organized criticality in the realm of economics.[43] We can view economies as being in a critical state, in which the sand avalanches correspond to crashes and business bankruptcies. (The positive feedback here is the freeing up of funds and people to start new businesses.) There will then be large market fluctuations, and there will be intrinsically unpredictable aspects of economic change. Moreover, the simple models of idealized equilibrium economies that seem to pervade the study of economics all fail to capture the essence of the self-organization.

Music may be another unexpected example of a critical phenomenon. A few years ago Richard Voss and John Clarke, two American physicists, noticed that a very wide range of musical compositions, spanning both Western and non-Western cultures, display a characteristic $(1/f)$ spectrum of intensity variations with sound-wave frequency, f, averaged over long periods.[44] In retrospect, one recognizes that the $1/f$ spectral pattern shared by all these types of music—from Beethoven to the Beatles—is one exhibiting a self-organized critical state. This is perhaps because we find the presence of patterns on all time intervals, which characterize this state, to be the most appealing, combining an optimal mixture of novelty and structure. Moreover, by being near a critical state, these patterns are sensitive to the nuances of performance in unpredictable ways. This provides a new insight into the appeal and freshness of musical performance, especially the nuances of timing, showing why we might expect the critical state to be the most alluring to the human mind.

The lesson we learn from the intriguing example of the sand pile is that, for this class of organized complex phenomena, there are unpredictable aspects of the critical state. As we learn more about the possible forms of complexity—and I believe that we shall find similar paradigms which characterize other equivalent classes of complexity—we may discover that the confluence of unpredictability and complexity is a very general one. We used to think that instability was a sign that something would not occur in Nature, or at best be very short-lived, like a needle balanced vertically on its point. But the sand pile shows us that many unstable events can come together to sustain a complicated manifestation of long-lived order. Life feels a little like this at times! And we don't have to look far to find other natural examples. The turbulent flow of the waterfall looks like this: each little eddy is chaotically unpredictable, yet it helps to sustain a flow of energy from large scales down to the smallest, which maintains the stability of the overall flow. Ultimately, we might even find that there is something of this criticality at the heart of our conscious minds as the brain self-organizes itself into a critical state. The firing of neurones one by one, triggering activity in other groups of neurones, has much in common with avalanches of sand. The functionality of the brain develops with time and might well approach a critical state. In such a state it would be most extensive and maximally sensitive to small changes. This type of susceptibility seems a very desirable feature of the system and may have some subtle link to the emergence of consciousness.

We have begun to understand some of the complex organized structures that we see around us. Eventually, we might hope that a fuller understanding of these structures and systems will allow us to produce them to order for specific purposes. Advanced civilizations with the ability to engineer optimal criticality into themselves, their technologies, and their environments, will be very different from ourselves. They will have decided to live with a high level of unpredictability. That unpredictability will make their futures a constant surprise in many ways, but they will know that the novelty that signifies critical efficiency can never be removed. In is a mark of the ultimate complexity that Nature can offer.

If we look at Fig. 5.6, we can see the range of complex structures that we have discovered or constructed. So far, they inhabit the middle range of sizes, above the atomic but below the astronomical. There is no reason why the examples of organized complexity should be limited in number. There may be an unending population of structures which conscious beings can engineer, or which Nature can produce under suitable natural conditions. It is in this realm of the complex that advanced civilizations have real scope to outstrip us. The study of the world of elementary particles might well be unexpectedly close to its end; the study of the astronomical universe might have very limited technological application; but the realm of the complex is immediately useful in countless ways. It offers a route to understanding life and consciousness, and of fabricating other examples of these fantastic processes. The number of complex structures will grow very rapidly with the number of permutations of the connections that can be made between different states. Our technology will have a hard job making a significant impact on this never-ending world of possibilities.

Demons: counting the cost

People who confuse science with technology tend to become confused about limits . . .
they imagine that new knowledge always means new know-how; some even
imagine that knowing everything would let us do anything.

ERIC DREXLER[45]

The nineteenth century was the period of industrial revolution. Scientists studied the efficiency of machines of all sorts, and gradually built up an understanding of the rules that govern the conservation and utilization of energy. The laws of thermodynamics were one of the results of these investigations. The most famous is the 'Second Law', which states that the entropy of a closed system can never decrease. In practice, this means that, even though energy is conserved in physical processes, it is degraded into less ordered, and less useful, forms, which are said to possess higher 'entropy'. There are so many more ways for a system to pass from order to disorder than there are ways for it to pass from disorder to order, that we habitually observe closed systems becoming steadily more disordered.[46] This principle is thus a statistical one; it is not a law of Nature of the same sort as the law of gravitation. However, it is of great importance for the consideration of what is technologically possible. It gives scientific precision to the idea that you can't get something for nothing—indeed, it means that you cannot even break even. Later, a succession of scientists explored the connection between entropy and information gain or loss. If we are to obtain information about the state of a system, then there is a cost. Work must be performed. The second law of thermodynamics allows us to evaluate the cost of acquiring information. Our path towards an understanding of these interconnections began with an almost frivolous suggestion by one of the greatest scientists of the nineteenth century.

Until the last decade of the nineteenth century, the laws of Nature had been about as impersonal as you could get. The observer and the observed were kept in complete isolation from one another. This was entirely in the Cartesian spirit of scientific investigation, which ignored any effect of the act of observation on the data being obtained. Science was like birdwatching from a perfect hide. In 1871, James Clerk Maxwell, the greatest British physicist since Newton, envisaged, for the first time, a situation in which a human-like intelligence might have to be accommodated by the laws of physics. He invited the readers to consider that

> if we conceive of a being whose faculties are so sharpened that he can follow every molecule in its course, such a being, whose attributes are still as essentially finite as our own, would be able to do what is at present impossible to us. For we have seen that the molecules in a vessel full of air at uniform temperature are moving with velocities by no means uniform, though the mean velocity of any great number of them, arbitrarily selected, as almost exactly uniform. Now let us suppose that such a vessel is divided into two portions, A and B, by a division in which there is a small hole, and that a being, who can see the individual molecules, opens and closes this hole, so as to allow only the slower ones to pass from B to A. He will thus, without expenditure of work, raise the temperature of B and lower that of A, in contradiction to the second law of thermodynamics.[47]

Maxwell was proposing that a 'sorting demon', as it became known,[48] could identify the faster molecules in a gas and steer them into one portion of a container, while the slower ones would be steered into the other portion, as illustrated in Fig. 5.8. (This is not unlike a nightclub whose doormen admit only 'attractive' customers.) The result would be a temperature difference between the two portions which could be used to drive an engine. This change of the system from uniform temperature and high entropy into one of two tem-

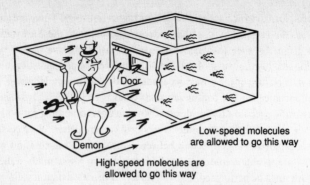

Fig. 5.8 Maxwell's Sorting Demon. The Demon was imagined to be able to distinguish fast and slow molecules and sort them into separate halves of the chamber by operating a door, so producing a temperature difference which could drive an engine.

peratures and lower entropy was manifestly a direct violation of the second law. Or was it?

The weak link in the demon's enterprise was identified by Leo Szilard in 1929,[49] and subsequently scrutinized in ever-greater detail by a host of others ever since. The problem is that the demon must distinguish between the fast- and slow-moving molecules, sort them into different regions, and then be ready to do this over again. In order to do these things, the demon must interact with the molecules in some way, say by shining a light on them and observing the colour of the reflected light. The work he must perform in order to discriminate between the fast and slow molecules and then destroy this information in order to repeat the operation from a clean start always outweighs the work that can be performed by exploiting the temperature difference created by the process of sorting.[50] Maxwell's demon is thus exorcized. It is not possible for him to create a violation of the second law of thermodynamics, any more than it is possible to show a profit at roulette by always betting on all the numbers. The cost of such a long-term strategy always outweighs the possible benefits.

These investigations have revealed that there is an absolute minimum amount of energy required to process a single bit of information. This minimum is determined by the second law of thermodynamics. Moreover, the physicist Jacob Bekenstein has discovered that in very general circumstances one can determine the maximum number of bits of information that can be stored in a region of given volume.[51]

We could classify civilizations by their ability to get as close as possible to the fundamental limits imposed by these restrictions, so as to create or harness systems at particular levels of complexity or with particular levels of information content. This quest has some very specific aspects; for example, the development of computers of ever greater size and processing speed. This development can be seen to proceed at two levels: there is the increase in the power of individual machines obtained by the optimization of their internal network of interconnections; but there is also a growth in their collective power produced by the networking of different computers. The Internet is the most familiar manifestation of this extension, but we could regard all non-local systems for information spread and retrieval, such as an international telephone system, as examples of this type. From a minimalist perspective, it is possible to classify all technological enterprises in terms of the amount of information needed to specify the structure completely and the rate at which that information needs to be changed in order for the system to change. In this way, we see that a thermometer is simpler (that is, it requires less information for its complete specification) than a desktop computer. The growth of a civilization's ability to store and process information has at least two quite different facets. On the one hand, there needs to be a growing ability to deal with things that become large and complicated; on the other hand, there is the need to compress information storage into smaller and smaller volumes of space. This storage compression takes place within the context of some hardware architecture, and so is intimately linked to the development of nanotechnologies.

These discoveries have taught us that information is a commodity. It takes effort to acquire it. As we have studied the physics of computation in greater detail, we have begun to appreciate the limits that are imposed upon technology and computing by the laws of thermodynamics and quantum physics. Any form of microscopic nanotechnological development will eventually encounter these fundamental limits. The most interesting of these is the discovery by the Nobel-prize-winning physicist Eugene Wigner of the limit on the size or mass of the smallest possible clock.[52] One might have expected the limit on the smallest clock size to be simply a limit imposed by the Uncertainty Principle of Heisenberg. However, a clock is a device which must be read repeatedly if it is to be useful. The limit on its minimum size turns out to much stronger than that imposed by the Uncertainty Principle of quantum mechanics, by a factor equal to the maximum running time of the clock divided by the smallest interval of time that you wish it to be able to resolve. Remarkably, it appears that the smallest microbacteria are quite close to achieving this limit, if one interprets their internal biocycle times as 'clocks'.[53] E. coli bacteria, at 0.01 micrometres, are barely one hundred times bigger than the quantum clock limit for structures with their mass. In the far future, one would expect these clock inequalities to place stronger limits than the Heisenberg Uncertainty Principle on the development of advanced nanotechnologies. All machines require coordination and timing over significant periods of time. Wigner's limit on minimum clock size restricts the size of timekeeping devices if they are to be robust enough to withstand the perturbations introduced by repeated observation. A failure of a living organism to do that would result in a breakdown of the synchronization and organization of its essential complexity. There is a limit to the miniaturization of complexity and technology.

We have seen that there is a three-way trade-off between time, energy, and information that is controlled by the limits on the amount of information that can be obtained with a given energy budget, the energy–time uncertainty principle, and the Wigner clock limit. We recognize the interplay of these three quantities in more familiar activities of everyday life as well. If we do things slowly then we use less energy than if we do them quickly. (Remember the calls in the USA to drive slower (at 55 m.p.h.) so as to make better use of the chemical energy in gasoline.) This link between speed and energy shows that reversible processes are more efficient (produce less waste heat) than irreversible processes. A perfectly reversible process goes infinitely slowly. It would take for ever to heat your house if you wanted the quality of the energy at the end of the heating process to be the same as at the beginning.

A Swiss physicist, Daniel Spreng, has schematized the interdependence of energy, time, and information as the triangle shown in Fig. 5.9.[54] Any two of the three attributes (energy, E, time, t, and information, I) can be traded in for the other two. Any point in the triangle represents a particular mixture of the three

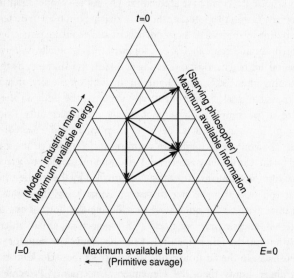

Fig. 5.9 Spreng's triangle, showing the symbiotic relationships of energy, time, and information. Each point in the triangle represents a possible mix of energy, time, and information necessary to complete a task. A change in any one of these three quantities is equivalent to a combination of changes in the other two.

ingredients needed to accomplish a given task. When lots of energy is available then you sit near the apex of the triangle, and as less and less energy is required you approach the bottom right-hand corner where the energy is zero. From the picture one can see how energy changes (or specific conservation measures) can be achieved by particular combinations of time changes and information changes. Near the corners of the triangle we find three distinct situations: at $E = 0$ there is the thoughtful philosopher, who takes very long periods of time and lots of information to accomplish his task; the primitive human ancestor, perhaps, lives near $I = 0$, and uses lots of time and energy doing things, because he lacks information about labour-saving devices; thirdly, near $t = 0$, there is the world of the modern (and future) technological society where lots of energy and information are employed to gets things done very quickly—the world of *Concord* and the Internet. In moving from one point in the triangle to another, the figure also shows what must be done to conserve energy. If we have lots of time we do not need much information because we can indulge in a haphazard trial-and-error search. But if time is expensive, then we need to know the fastest way to do things and that requires lots of information. Alvin Weinberg has argued that this means that time is likely to become, increasingly, our most important resource. The value of energy and information is, ultimately, that it gives us more freedom to allocate our time,

> In ... the Computer Age, I would suggest that the reorganisation of our use of time may be the most profound and lasting social effect of the extraordinary advances in the handling of information that have largely resulted from the work of ever more efficient computing machinery.[55]

Two types of future

Confident articles on the future seem to me, intellectually, the most disreputable of all forms of public utterance.

KENNETH CLARK[56]

Just as the Western societies in which most scientists operate tend to produce two strands of political opinion with right and left wing, liberal, or republican tendencies, so one finds a similar divide in eschatological prognostications. There are those who see the future of life in the Universe as a never-ending battle between competitors in the form of rival intelligences (who will ultimately include machines) and Nature itself. In contrast, there are others who see a future of enlightenment, cooperation, and harmonious equilibrium. In many ways, these two alternatives reflect two possible end-states which can be reached by any two populations engaged in a competition with one another. This problem has been extensively analysed by mathematicians under the heading of 'game theory'. A 'game' is a collection of alternative strategies in which two or more players may indulge (either consciously or unconsciously) and the pay-offs that accrue to them. For example, players could be individuals doing business and the strategies could be 'match rival bids', or 'undercut rival bids'. We might then ask which strategy is best in the long run by evaluating all the possible outcomes and their benefits and penalties. In general, one would like to know the best strategy to adopt in the sense that it is the best possible (or least risky) regardless of what your competitors do.

Over a long period of time we would like to know if competitions settle down to some final state of balance or whether conflicts keep escalating. One possible final state is for what is called an *evolutionarily stable strategy* (ESS) to be adopted by all players.[57] This strategy is stable in the sense that any player who deviated from it would be worse off. It is, however, always possible that no such ESS exists; it depends on the rules of the game. For example, if we take the old children's game of 'rock–scissors–paper' played between two players, then if both players have to pay an equal penalty whenever there is a draw (that is, when they both show the same hand) then the ESS for any player is to play a mixed strategy in which rock, scissors, and paper each appear with equal probability (one in three) in the long run. But if the rules are changed, so that both players are given an equal reward when they play the same hand, then there turns out to be no ESS at all.

Alternatively, players may find themselves in a *rat race* in which each escalates his or her response to the previous move by the other, as if they were in an arms race. Thus, two rival species in a bounded habitat might evolve over very long periods of time so that one grew sharper teeth while the other grew thicker armour. Typically, one expects games involving resources which are limited in some way to tend towards the adoption of an ESS with respect to the utilization of those limited resources, while those aspects which are not so constrained may become engaged in a rat race. If, for example, different varieties of tree in a dense forest are competing for light in the forest canopy, they could each try to grow taller than the other, but there is a constraint on this arms race because the strategy costs time and energy.

Most speculators on the ultimate fate of the scientific enterprise, and of scientists, can be placed in one of these two camps: the ESS or the rat race. One group of commentators sees the technological age as something that will ultimately be transcended by a race of cerebral beings who learn to counter their urge to expand their territory and to manipulate Nature, like the Overlords in Arthur C. Clarke's novel *Childhood's End*.[58] Only by halting technological advance will they be able to live within the bounds of their planetary system and remain in some measure of equilibrium with their environment. This prognosis is very much an extrapolation of serious concerns about the exhaustion of the

Earth's present resources. It is often predicted that these advanced beings would have to possess sophisticated altruistic and ethical principles, because these qualities appear to be necessary conditions for any ultra long-lived civilization to persist.[59] This scenario is quite consistent with the expectation that one consequence of ultra-advanced technology would probably be the enormous (or even indefinite) extension of individual lifetimes. This would lead to a slowing down in the evolution of diversity, perhaps even with long periods of self-imposed hibernation, and would result in a form of self-engineered ESS. This view is common among enthusiasts for extraterrestrial intelligence and those engaged actively in the search for it.[60] This is not surprising. Since the greatest possible pay-off from such searches would be contact with extremely advanced intelligent life forms, it is important to convince ourselves that their intentions towards us would be entirely honourable. If we believed otherwise, then our best strategy would be to develop effective smokescreens to hide the evidence of our existence, rather than to broadcast our presence (and lack of intelligence) over the interstellar radio spectrum. One astronomer who favours this scenario is Michael Papagiannis, of Boston University, who believes that idealistic advanced civilizations

> that manage to overcome their innate tendencies toward continuous material growth and replace them with non-material goals will be the ones to survive the crisis. As a result the entire Galaxy in a cosmically short period will become populated by stable, highly ethical and spiritual civilisations.[61]

The alternative picture sees survival becoming harder and harder for long-lived civilizations. They may even have had to regenerate their civilizations on several occasions, following disasters of war or impacts by comets and asteroids on their planets. Their behaviour may have evolved along quite unusual evolutionary tracks, and they may display some quite unexpected by-products of their evolution (music, mathematics, . . . , etc). One expects that the more advanced an intelligence becomes, so the more extensive, non-linear, and unpredictable will be the by-products of their intelligence. Biologists have good reason to believe that altruism is a strategy that is optimal in fairly general circumstances, and some altruistic behaviour can be selected for without the need to impose it by adopting ethical codes. However, the virtuous qualities preached and revered by many terrestrial religions cannot be explained by adaptive evolution alone. They advocate selfless acts which greatly outstrip the level of altruism and self-sacrifice that is optimal from the narrow evolutionary perspective.[62]

Scientists and futurists like Hans Moravec[63] or Olaf Stapleton[64] have seen scientific progress as a necessary outcome of competition between 'computers' or similar advanced intelligences. But we do not know whether competition is a phase of evolution which it ultimately pays to replace by cooperation. On Earth, we see a global move towards cooperation in many spheres imposed by economic restrictions. Perhaps the pattern will recur over larger and larger scales in the far distant future.

Is technological progress inevitable (or always desirable)?—a fable

> The future is a fabric of interlacing possibilities, some of which will gradually
> become probabilities, and a few which become inevitablities, but there are
> surprises sewn into the warp and the woof, which can tear it apart.
>
> ANNE RICE[65]

'You can't get there from here' is the answer you are tempted to give to the motorist lost in the one-way system, seeking directions to the other side of town. And so, when we envisage distant futures with everything better and faster than now, we must also wonder whether it is possible to get there from here and now. We have all experienced the 'improved service' that is worse than its predecessor—merely an improvement for the supplier who can provide a bit less for a lot less. Worse still, there is a form of progress that creates unavoidable adverse side-effects which only come to light too late. We know of many foods and medicines that have turned out to have serious side-effects. We have realized, too late, that industrialization has wrought huge changes to the balance of our climate and ecosystems. There is an obvious pattern. The more powerful and far-reaching the benefits of a technology can be, so the more serious are likely to be the by-products of its misuse or failure. The more structure that a technology can bring about from randomness, so the further its products depart from thermal equilibrium, and the harder it is to reverse the process that gave rise to them. As we project a future of increasing technological progress, we may face a future that is increasingly hazardous and susceptible to irreversible disaster.

It is hard for many people to imagine how progress can leave you worse off. A particularly intriguing example, using only psychological impulses that we recognize only too well, was created in 1951 by the science fiction writer Arthur C. Clarke. *Superiority*[66] is the story of a technologically super-advanced civilization that finds itself defeated in a space war by a technologically inferior force. The story is told by their demoralized and defeated Commander-in-Chief from the prison cell where he awaits sentence and punishment for incompetence. The plea entered by the defence counsel in mitigation is that it was the military's unquestioning faith in technological and scientific progress that led to their disastrous defeat by the inferior science of their enemies. In his defence, the Commander tells us the paradoxical story of his Starfleet's defeat.

At first, the Commander-in-Chief was confident that Starfleet's greater numbers and superior military science would bring a quick and easy victory over their enemies. But, although successful in the early battles, the margins of victory were a lot narrower than expected. Things could easily have gone against them. Shocked by this, a conference was called, and the new chief of weapons research, General Norden, took the floor to argue very persuasively that their weapon systems had evolved to a dead end. They were too conservative. It was no use just trying to perfect old weapons by making tiny improvements. New, more adventurous, high-tech systems were needed. Norden replaced the old guard of weapons scientists, and announced that he would accelerate the production of a new suite of weapons that were still under development.

The Commander and his team of generals were understandably nervous, but there was little they could do to influence their political masters after Norden's technicians successfully demonstrated a new weapon, the Sphere of Annihilation, just four weeks later. This device produced complete disintegration of all matter within several hundred metres of it. All existing missile guidance systems were altered to accommodate the spectacular new weapon. This did not go as smoothly as Norden had expected. Only the very largest missiles could cope with the load, and these in turn could be carried only by the heaviest spaceships in the Starfleet. However, Norden was such a charismatic technical genius that no one worried about these limitations at the time. Everyone was confident that victory was inevitable.

Soon afterwards, things began to take a turn for the worse. One of Starfleet's own ships disappeared after a Sphere triggered itself immediately after launch. Morale plummeted, and relations between Starfleet's crews and Norden's scientists soured. Norden responded by announcing a tenfold increase in the range of the sphere of destruction, but more changes had to be made to all the launch systems. Still, everyone convinced themselves that the improved technology would be worth the wait. Meanwhile, though, the opposition simply beat back the Commander's forces, emboldened by the absence of any attacks against them. Despite their superior numbers, the home forces could barely hold off these incursions because so many of their weapon systems were out of action while being upgraded. Several small bases on the outskirts of the Empire were lost completely.

The enemy had been frantically building more and more of its old-fashioned low-tech ships and weapons: soon, they had established a significant numerical advantage. Norden argued that quantity was no match for quality. He seemed to be right. Although there had been teething troubles with the Sphere, when it worked it destroyed many enemy ships. Still, territory was steadily being lost, and the enemy was getting more daring. Norden began to be strongly criticized by the Starship commanders. He responded by unveiling a new top-secret weapon, the Battle Analyser. This was an intelligent computer system, designed to manage the complexities of battle automatically. Unfortunately, bad luck struck. The first system, together with five thousand of the best technicians, was placed on board a ship that struck a wayward space mine. The loss was total. Norden faced disgrace, but he responded with a weapon so fantastic, so powerful, so unexpected that the military commanders could hardly believe it. The Exponential Field, Norden explained to his stunned audience, could send a portion of space off to infinity. Nothing could reach a Starship carrying the Field. Even when completely surrounded by enemy ships it was unassailable. It could disappear when attacked; materialize without warning next to an enemy ship, destroy it, and disappear again. This was the ultimate secret weapon. The equipment needed was very simple and inexpensive. All ships were re-equipped with the Field, putting them temporarily out of action yet again, but confidence was returning. Advanced technology had paid off . . . or so it seemed.

At first, the trials went well. The Starship commanders were amazed at the way they could hop around the Universe at will. But then some minor problems appeared. Nothing serious; just communications circuits not working properly. They returned to base to get these sorted out, but the enemy suddenly launched a major offensive and the ships carrying the Field weapon had to be relaunched into battle before the repairs could be fully tested. On seeing the vast enemy force, the whole Starfleet switched on the Field, and promptly disappeared into hyperspace. They followed their instructions to the letter, charting their precise return in groups, to outnumber and surprise the enemy ships one by one. They never knew what went wrong. Disaster struck when they returned, each to a different place than the one they had planned. Some, right in the middle of the enemy formation, were immediately destroyed. Others found themselves lost on the other side of the Galaxy. Worse of all, none of the ships could make contact with any others. The communications equipment seemed to be working perfectly though, and so each Starship's commander began to believe that his ship was the sole survivor.

Only later, after total defeat and the capture of the home planet by the enemy, did the awful truth emerge. Whenever the Field was switched on, it caused a hyperspatial distortion of the ship and all its components as they were whipped off towards infinity. When the Field was reversed the distortions were reversed—but never perfectly. There were always tiny errors. Entropy had increased. Things never returned to how they were—or where they were— relative to their immediate neighbours. At first, the little mismatches thus created in the electronic systems were too small to have any effect at all. But they were cumulative: after any Starship had used and reversed the Field a few times its components and electrical circuits started to drift away from the specification of those on other ships in the fleet. Communication frequencies and codes started to drift out of synchronization, and some delicate high-tech systems just wouldn't work at all. Things got worse and worse, culminating in total chaos. The enemy was attacking with thousands of its primitive ships and outdated weapons. Every time a ship used the Field to flee from the attackers, its equipment was further distorted. Eventually, nothing would work; every ship was isolated. They were doomed. All the Commander-in-Chief could do was surrender, his fleet defeated by its own more advanced science.

The lessons of this poignant story are obvious. The drive for progress, accompanied by a declining knowledge of science by the end-users and political ringmasters of science, can lead to irreversible disaster. The more sophisticated and powerful a technological system becomes, so the more susceptible it is likely to be to breakdowns and subtle malfunctions. Similarly, the more far-reaching will be the consequences of those breakdowns.

Progress makes existence more complicated and disasters more devastating. This does not mean that we should respond negatively, by avoiding progress, preaching always a message of paranoia about the dangers of technology. There will no doubt be particular technological developments that we shall want to scotch because they create unacceptable risks, but our general response should be to make sure that our analyses of risk, and our standards of safety, progress hand in hand with the technology. Electricity is dangerous. This does not mean that we cease to use it, or veto further development of its applications. Instead, we try to introduce strict standards of practice to ensure safety.

Summary

The difficult is that which can be done immediately; the Impossible that which takes a little longer

GEORGE SANTAYANA

Someone once said the acid test of all scientific progress is whether it allows us to build better machines. This view is provoked by the position that we occupy in the spectrum of sizes of natural things. We are far bigger than the atoms and far smaller than the stars. We must create artificial senses if we are probe the worlds of the large and small, understand environments that display extremes of temperature and density, or come to terms with overwhelming complexities. We have found that the path to understanding the deep structure of the Universe, its laws and complex states, leads us to explore conditions far removed from those which were familiar to our ancestors. The limits to what we can ultimately discover are likely to be imposed by limits of technology rather than by limited imagination. Already, our most successful theories of Nature's forces make precise predictions about the workings of the Universe under conditions that, at present, we cannot remotely approach by direct experiment. Indeed, in order to discover whether our version of Nature's laws is the correct one it looks as if it is necessary to investigate what happens when matter is subjected to temperatures more than 10^{15} (1,000,000,000,000,000) times as great as those achievable in our most powerful terrestrial experiments. It is unlikely that direct experiments of that sort will ever be possible.

Unfortunately, our technological powers are confronted by a variety of limits. Some are financial and practical. Democracies will not be willing to devote large fractions of their GNP to activities which offer no immediate return when society is confronted with serious environmental or medical problems that require scientific solutions. These limits will recede only if entirely new ways are found to generate energy. But there are yet deeper limits to experimental inquiry.

We have speculated about the steps that civilizations might take as they ascend to master the realms of the large and the small. Ultimately, these advances will have to come to terms with the limits that Nature imposes on how fast we can transmit information, how small we can ensure accurate time-keeping, how much energy must be expended to gain information, how close to criticality are the complex systems that we see, and how sensitive is our technology to errors and the chaotic amplification of uncertainties.

The development of technology, and the ability to test the theories that we have about the behaviour of matter under extreme conditions, require us to manipulate matter, energy, and information over scales that are increasingly divorced from those of our everyday experience. Intriguingly, the decisive features of the laws of Nature appear to be manifested in these extreme environments. By delving into them we are not merely seeking completeness for its own sake: the behaviour of matter at ultra-high temperatures is the crux of its most basic character. One of the ways in which we could sidestep these limits on our ability to create high energies is by using astronomical observations. Our universe is expanding and appears to have experienced extremes of temperature and energy during its early stages.[67] If its early history left behind observable relics of its fiery birth, then they might provide a new window on the behaviour of matter at the highest imaginable energies. It is to this cosmological story that our attention now turns.

CHAPTER 6

Cosmological limits

*I do not know what, if anything, the Universe has in its mind, but I am quite, quite
sure that, whatever it has in its mind, it is not at all like what we have in ours.
And, considering what most of us have in ours, it is just as well.*

RALPH ESTLING[1]

The last horizon

*One of the problems has to do with the speed of light and the difficulties involved in trying
to exceed it. You can't. Nothing travels faster than the speed of light with the possible
exception of bad news, which obeys its own special laws.*

DOUGLAS ADAMS[2]

Cosmology is a special study: its subject is unique, its object is unique, and its
means are unique. No branch of science extrapolates so far into the unknown,
and no line of human inquiry is more at risk from limits of all sorts. The
cosmologist must overcome the technological challenge of seeing faint objects
at great distances, and succeed without many of the weapons in the scientist's
armoury.

Unfortunately, we cannot experiment on the Universe; we can only look at
what it has to offer. When we look at astronomical objects, like stars and planets,
we can take the outsider's view, but when it comes to the Universe as a whole we
cannot get outside it: we are part of the system we are trying to describe. This
creates some peculiar problems that the scientific method was never designed to
deal with.

In the past decade there has been huge progress in our knowledge of the
astronomical universe. Technological ingenuity has provided us with light
detectors of unprecedented sensitivity. Space agencies have launched astronomical satellites able to look at the Universe across the whole electromagnetic
spectrum. The highlight of this programme—the launch of the Hubble Space
Telescope (and the subsequent application of COSTAR, its corrective optics
package)—has enabled us to look at planets, stars, and galaxies with astonishing
resolution.[3] The fuzziness created by the scattering of light by the molecules in
the Earth's atmosphere, the same scattering that makes the stars twinkle, has
been removed. Familiar objects have suddenly been brought into a focus so

sharp that all manner of unsuspected structure has emerged, shedding new light on how stars and galaxies are formed. But, most dramatically, we have been able to see things that are further away than we have ever seen before (Fig. 6.1). Time and again, these images have been emblazoned across the world's media for the admiration of the public, while professional astronomers have been racing to keep up with the flood of new information.

When we look at distant galaxies with an instrument like the Hubble Space

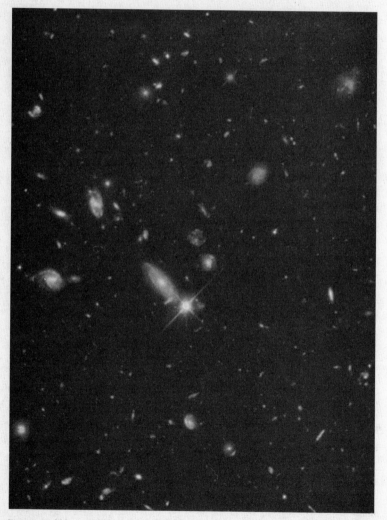

Fig. 6.1 Hubble Space Telescope 'Deep field': the deepest exposure ever taken of the Universe.

Telescope, we must recall the most striking fact about our view of the Universe. Light moves with a finite speed, and so when we 'see' a distant galaxy now, we see it as it *was* when the light left it, not as it *is* today. The Universe provides us with the simplest form of time machine, one that allows us to see the distant past just by looking. The most distant objects that we can see are billions of light years away: their light has taken billions of years to arrive. They are younger versions of the mature galaxies, like our own Milky Way, that we see near by. Some people worry about how we can know what the Universe was like billions of years ago. Actually, the real problem is knowing what it is like *now*.

These exciting developments have inspired many popular accounts of the present state of the Universe,[4] together with new theories about its beginning and possible future state. These extrapolations of our present observations are possible because we have a theory[5] of how the Universe changes with time. Einstein's theory of general relativity is the basic tool for these studies. It supplies equations which tell us how any universe containing matter and radiation will change with time under the influence of gravity. Unlike Newton's theory, it can deal with motions at, or near, the speed of light, and also with very strong gravitational fields. Einstein's equations allow us to reconstruct the history of the Universe and so discover the sort of past which can give rise to the present. This presents a special problem. The Universe is expanding; so looking backwards in time requires us to contemplate times when the Universe was hotter and denser than it is today. As we look backwards, at first there will be no galaxies, then no stars, then no molecules or atoms, then no nuclear elements, and eventually no protons and neutrons: just a soup of the most elementary particles of matter and radiation. So far, we have a good understanding of what the composition of the Universe would be like from the point when it was only about one second old. But when we try to probe even further back, we need to have a fuller understanding of the elementary particles of matter than we have at present. The conditions that we have to deal with are more extreme than any we can create by artificial means on Earth in particle colliders and accelerators, and so our reconstruction of the Universe's past becomes uncertain in crucial ways.

At present, the expansion of the Universe appears to be proceeding in an extremely uniform way. It goes at the same rate in every direction to a precision better than one part in a million. Observations of radio waves left over from when the Universe was about a million years old show that the Universe was extremely uniform from place to place as well. Only later, when the Universe was billions of years old, did matter become aggregated, non-uniformly, into luminous collections of stars and galaxies. Consequently, cosmologists take the simplest possibility, and start from the assumption that the Universe has always been uniform, with just very small irregularities in the overall uniformity of expansion. Although small, those non-uniformities are rather important. Places that contain more matter than average pull even more matter towards them, at the expense of the sparser regions, like a form of cosmic 'Matthew Principle' ('For whosoever hath, to him shall be given, . . . but whosoever hath not, from him shall be taken away even that he hath.'[6]). In time, the denser-than-average regions turn into galaxies, stars, and people.

One of the tasks of cosmologists is to come up with something better than this simplified broad-brush story: for example, to show that the present state of the Universe is an inevitable consequence of what Einstein's equations say about expanding universes, or of the behaviour of matter at very high densities. We would like to build realistic computer simulations of the entire sequence of events that transform regions of greater than average density into structures that look like real galaxies of stars, gas, dust, and other non-luminous material.

As indicated above, the Universe is assumed, as a first approximation, to be the same everywhere and to expand at the same rate in every direction. This expansion can then be described by a single quantity, the scale factor, which is a measure of the separation between any two reference points. Its actual value has no physical meaning; all that matters is the ratio of its values at two different times. This tells how much expansion of the Universe has occurred. The scale factor (which is sometimes referred to, rather inaccurately, as the 'radius of the universe') can vary in time in two distinct ways, as shown in Fig. 6.2. It can increase for ever (an 'open' universe), or it can expand to a maximum and then decrease (sometimes called a 'closed' universe). In between, there is a compromise universe (sometimes called 'flat' or 'critical'), which just manages to

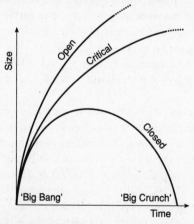

Fig. 6.2 The possible variation of the distance with time in expanding universes. There are three characteristic types: 'closed', 'open', and 'critical'.

expand fast enough to continue expanding for ever. It is the cosmic dividing line between the open and closed universes.

In popular expositions of cosmology for non-specialists (and even in some for specialists), a number of simplifying assumptions are usually made which obscure the fundamental limitations on our ability to answer familiar questions about the Universe.[7] We shall see that many of the questions that popular accounts of cosmology raise, and sometimes even confidently answer, appear to unanswerable. Answers can be given only because some untestable assumptions have been smuggled in to simplify the problem, or to rule many possibilities out of court from the outset. As a result, there are limits to what we can know about the Universe. Those limits cut across all the major unsolved problems of cosmology.

One important thing to notice about our simple picture of the expanding Universe is the impact of a simplifying assumption that is already implicit in the drawing of Fig. 6.2: that the Universe is the same everywhere. This means that we can talk about the expansion of the Universe in terms of the *single* measure of its size, rather than a collection of them—one for each location in the Universe. We can easily fall into the habit of thinking that our observations characterize the entire Universe rather than just a part of it—that part which we can see. Let us look at this problem more closely.

First, we must distinguish between two meanings of 'universe'. There is the *Universe* with a capital U—that is, everything there is. This may be finite, or it may be infinite. In addition, there is also something smaller that we call the *visible universe*. This is a spherical region centred on us, from within which light has had time to reach us since the Universe began. Since light travels at a finite

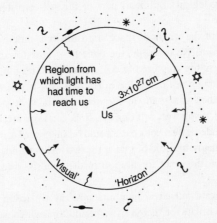

Fig. 6.3 The visible universe is defined to be a finite spherical region of radius equal to the distance that light can travel in the time since the expansion began.

speed in vacuum (and nothing travels faster), the visible universe has a finite size. It constitutes all that we could possibly see of the Universe, in principle, today, with perfect measuring instruments of unlimited sensitivity. The boundary of our visible universe is called our *horizon*. It defines the boundary of observational science and its size increases steadily with the passage of time, reflecting the fact that more and more light has time to reach us.[8]

The first lesson we draw from this simple observation is that astronomy can only tell us about the structure of the *visible* universe. We can know nothing of what lies beyond our horizon. So, while we might be able to say whether our visible universe has certain properties, we can say nothing about the properties of the Universe as a whole unless we smuggle in an assumption that the Universe beyond our horizon is the same, or approximately the same, in nature as the visible universe within our horizon. This prevents us from making any testable statements about the initial structure, or the origin, of the whole Universe.

If the Universe is finite, then the visible universe will always be a finite fraction of the whole. By contrast, if the Universe is infinite in size, then our observations will only ever sample an infinitesimal portion of the whole. We shall never know for certain which of these situations is ours. Einstein's equations, which tell us what universes there can be, allow both infinite and finite universes.[9] It is possible that some future development in the study of how to unify gravity and quantum physics will produce a strong result of the form that the Universe must be finite or a theory of quantum gravity cannot exist, or that it must be infinite to avoid some other deep internal inconsistency. A theoretical result of this sort might be very persuasive to cosmologists. Although it would not be observational evidence for the finiteness, or otherwise, of the Universe, it would be seen as a strong logical argument, part of the self-consistency of quantum theory.

Let us represent the whole of space and time by a simple picture, called a *space–time diagram*. We represent the passage of time towards the future as the vertical scale of the graph in Fig. 6.4, and all the three dimensions of space are portrayed along one line, as the horizontal axis. If you remained at one place in space, then your path in the diagram would be on a vertical line moving upwards. If you were orbiting in a circle (and you are, when the Earth's motions are included), then your path would be an upward spiral. The path of a light ray in this diagram would be one of the two inclined lines (one for motion from left to right, the other for motion in the opposite direction) shown in Fig. 6.5.

Let us now locate ourselves 'here and now' at a place in the space–time diagram. We can isolate the region of space and time which we can investigate by receiving light rays or other, more slowly moving, signals. This region consists of the shaded cone, and is called *our past light-cone*. When astronomers

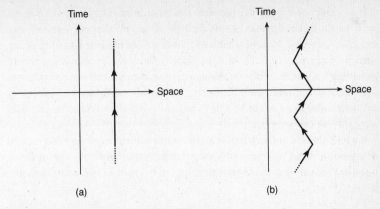

Fig. 6.4 The space–time path of a point (a) remaining at the one location as time passes; (b) moving back and forth in space as time passes.

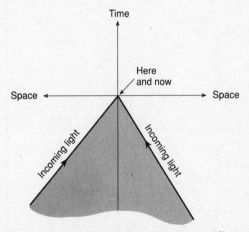

Fig. 6.5 The paths of light rays received here and now, in a space–time diagram.

receive any type of light ray, whether it be from optical, X-ray, infrared, ultra-violet, or radio sources, these give information about the structure of the edge of the cone. The further away, and hence the earlier, that they originate, so the greater the distance down the surface of cone that they permit us to probe.

When we collect massive particles, like cosmic rays or meteorites, which travel more slowly than the speed of light, then they reveal something about the inside of our past light-cone. In fact, when we collect fossils, or study the interior of the Earth, we are also obtaining information about the Universe inside our past light-cone.

If we mark out the region from which we have direct information, it is really quite small. Beyond the surface of the light-cone—outside our horizon—we can know nothing. Most of what we know describes the past structure of the surface of our past cone. If we think that we know the mathematical theory governing the way the Universe changes in time, then we can use it to calculate inwards and outwards away from the surface of the light-cone. When we calculate inwards we might be able to test our predictions. When we calculate outwards, we cannot.

If a scientist or a philosopher wants to assume something about the structure of a point in the Universe that is today far from our light-cone (point *P*, for instance) then an untestable extrapolation can be made in two ways, as shown in Fig. 6.6.

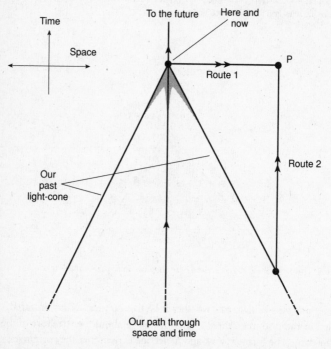

Fig. 6.6 Two ways of extrapolating from the observable universe to the unobservable Universe. A space–time diagram showing our past light-cone, from the edge and interior of which all our observational information comes. If we want to say anything about an unobservable point, *A*, then we usually do so by extrapolating conditions from places in space–time about which we have data. We can take Route 1, assuming that conditions elsewhere are as they are here at our location today, or we can take Route 2, assuming that the history of *A* back to its intersection with our past is similar to our history.

Route 1 assumes that the Universe is the same (or approximately the same) everywhere, at the same moment of time. Route 2 assumes that the present state of P can be found by running forward from the place where its history cuts our past light-cone; that is, we assume that things outside our light-cone change in the same way as they did during our astronomical history.

The observations made by NASA's Cosmic Background Explorer (COBE) satellite tell us something about the structure of our past light-cone when the visible universe was about a thousand times smaller than it is today, about three hundred thousand years after it began expanding. This is the time when the universe had expanded enough for the radiation to cool and cease interacting with electrons. Then it will fly freely towards us through space and time. COBE reveals that the visible universe was extremely uniform from one direction to another at that time. However, before that time, the Universe is opaque to photons. The scattering of photons by electrons prevents us from seeing back any further. If it ever becomes possible to detect neutrinos from the early universe, we shall then be able to see back to just one second after the expansion began, when the region comprising our present visible universe was ten billion times smaller than today. Before then, the universe will be opaque to neutrinos as well. Our only hope for direct observation will be by means of gravitational radiation. We might in principle see all the way back to when the universe was 10^{32} times smaller than now. Technologically, these are challenging problems for the far future.

Fortunately, we can still learn things about the Universe when the expansion was one second old with today's technology, by observing the abundances of the lightest chemical elements in the universe. Elements like helium and lithium, together with isotopes of hydrogen like deuterium, are produced by nuclear reactions at the end of a sensitive process that began when the universe was one second old, and ended when it was a few minutes old. By comparing our observations of these elements with the predictions made from our model of what the universe must have been like when it was one second old, we can test the model. Unfortunately, we have not been able to play this game for the period before one second. So far, we have not found any 'fossils' left over from the first second of the universe's history. We can, however, turn the game round. There are many models of what the universe was like during the first second of its history, incorporating different theories of the behaviour of elementary particles of matter at high energy. Some of them can be ruled out because they predict things that we do not see today.

How worried should we be about this absolute limit on our ability to determine the structure of the Universe? Before about 1980, the difference between the visible universe and the whole Universe was ignored because cosmologists could find no positive reason to believe that the Universe should be very different in structure beyond our horizon. The distinction seemed rather anti-Copernican: maintaining that *our* visible universe was special or atypical in some way. During the 1980s that situation has steadily changed. A new version of the Big Bang theory gave reason to expect the Universe to be very different inside and outside our horizon.

Inflation—still crazy after all these years

'How long will these lectures continue?' asked President Gilman one day of Lord Rayleigh, while walking away from the lecture-theatre. 'I don't know', was the reply; 'I suppose they will end some time, but I confess I see no reason why they should.'

SILVANUS THOMPSON[10]

Since 1980, the preferred theory of the very early Universe has included a historical interlude called 'inflation'. It adds a slight gloss to the simple picture of an expanding universe. But this gloss has huge implications. The standard picture of the expanding universe, which has been with us since the 1920s, has a particular property: the expansion is decelerating. No matter whether the universe is destined to expand for ever, or to collapse back in on itself towards a Big Crunch, the expansion is always being decelerated by the gravitational attraction exerted by all the material in the Universe. The deceleration is simply a consequence of the attractive character of the force of gravity.

It had always been assumed that gravity would ensure that matter and energy would attract other forms of matter and energy. But in the 1970s particle physicists began to find that their theories of how matter behaved at high temperatures contained a collection of matter fields, called *scalar fields*, whose gravitational effect upon each other could be repulsive. If those fields were to become the largest contributors to the density of the universe at some stage in its early history, then the deceleration of the universe would be replaced by a surge of acceleration. Remarkably, it appeared that if scalar fields do exist, then they invariably come to be the most influential constituent of the universe, and their influence ceases only when they decay into ordinary matter and radiation.

The inflationary universe theory is simply that a brief period of accelerated expansion occurred in the very early history of the Universe, perhaps because one of these ubiquitous scalar fields came to dominate the density of matter in the universe. The field then needs to decay quite rapidly. When it does so, the expansion resumes its usual decelerating expansion (Fig. 6.7). This sounds innocuous, but a very short period of accelerated expansion can solve many long-standing cosmological problems.

The first consequence of a short period of accelerated expansion in the past is that it enables us to understand why our visible universe is expanding so close to

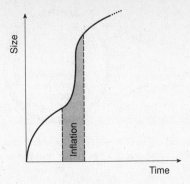

Fig. 6.7 The variation of distances with time in an inflationary universe. The expansion is 'inflated' by a brief period of early acceleration.

the critical divide that separates open universes from closed ones. The fact that we are still so close to this divide, after about fifteen billion years of expansion, is quite astonishing. Since any deviation from lying precisely on the critical divide grows steadily with the passage of time, the expansion when it started must have been extraordinary close to the divide in order to remain so close today—so close that we still do not know on which side of the divide we lie. (We cannot lie *exactly* on it.[11]) But the tendency of the expansion to veer away from the critical divide is just another consequence of the attractiveness of the gravitational force. If gravity is repulsive and the expansion accelerates, then, while it lasts, the acceleration will drive the expansion ever closer to the critical divide. If inflation lasted long enough, it would explain why our visible universe is still so close to the critical divide.[12]

Another by-product of a short bout of cosmic acceleration is that any irregularities in the expansion of the universe get ironed out and the expansion very quickly goes at the same rate in every direction, just as we see today. This offers an explanation for a property of the expansion of the universe that has always struck cosmologists as mysterious and unlikely—because there are so many more ways to expand in different ways in different directions.

Third, the visible universe around us today will have expanded from a region that is much *smaller* than it would have originated from had the expansion always decelerated, as in the conventional Big Bang theory. The smallness of our inflationary beginnings has the nice feature of offering an explanation both for the high degree of uniformity that exists in the overall expansion of the Universe, and for the very small non-uniformities seen by the *COBE* satellite. These are the seeds that subsequently develop into galaxies and clusters.

If the Universe accelerates, then the whole of our visible universe can arise

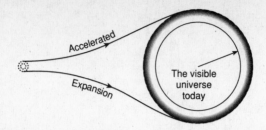

Fig. 6.8 The inflation of a small region of the very early universe. Inflation expands a region that is small enough for light to traverse to a size greater than that of the visible universe today.

from the expansion of a region that is small enough for light signals to traverse it at very early times (Fig. 6.8). This light traversal enables conditions within that primordial region to be kept smooth. Any irregularities get smoothed out very quickly. In the old, non-inflationary Big Bang theory the situation was very different. Our visible universe had to emerge from a region vastly bigger than one that light rays could coordinate and smooth. It was therefore a complete mystery why our visible universe looks so similar in every direction on the sky to within one part in 100,000, as observations have shown. One part of the universe would not have had time to receive light rays from another part far away.

The tiny region which grew into our visible universe could not have started out perfectly smooth. That is impossible. There must always be some tiny level of random fluctuation present: Heisenberg's Uncertainty Principle requires it. Remarkably, a period of inflation stretches any fluctuations to very large astronomical scales, where they appear to have been seen by NASA's *COBE* satellite.[13] In the next few years, they will be subjected to minute scrutiny by two more satellites full of instruments, which are currently being prepared for launch. If inflation occurred, the signals they see should have very particular forms. So far, the data taken by *COBE* over four years are in very good agreement with the predictions, but the really decisive features of the observable signal cannot be seen by *COBE*. The two new satellites, *MAP* (to be launched in 2000) and the *Planck Surveyor* (to be launched in 2005), will decide this question.

Cosmologists have always faced a dilemma about the beginning of the Universe. If the present structure of the Universe depends in some way on the way that the Universe began (and whether it *did* have a beginning), then our astronomical observations might tell us something about the initial state of our visible part of the Universe. But there is a downside to this. It means that any 'explanation' of why the Universe is as it is today would boil down to a statement about why it was as it was, and ultimately to a statement about its structure in the beginning. Since we do not expect that knowledge of the initial state is likely to be forthcoming, there has always been a desire for a different type of cosmological explanation. Suppose that it could be shown that the gross features of the observable universe arise no matter how it began, as long as the expansion goes on for long enough. If so, we could explain the present structure of the Universe without any need for a detailed understanding of what it was like in the beginning.

In 1967, the isotropy of the microwave background radiation was first discovered. It was found to be constant in temperature around the sky to within less than one part in a thousand. This remarkable uniformity challenged cosmologists for an explanation. They had previously been struggling to come up with an explanation for the small irregularities that grow to become fully fledged galaxies. Suddenly, they realized that it was the underlying uniformity (rather than the little lumps and bumps) that was most in need of explanation.

An American cosmologist, Charles Misner, proposed that it might be possible to show that if the universe began in a highly irregular and anisotropic state, then frictional processes would erase all the irregularities early on.[14] The universe would be left to expand towards the symmetrical state that we see today, provided that it expanded for long enough.

The general idea, of showing that universes that began in a chaotically irregular state would eventually smooth themselves out, was dubbed the 'chaotic cosmology programme'. This ambitious project foundered. There were too many awkward forms of irregularity that would not go away fast enough, and some forms that would not go away at all. Moreover, getting rid of the irregularity by frictional dissipation produced far more waste heat than we find in the universe today.[15]

The most important thing to appreciate about the chaotic cosmology programme is the following. If we can find an explanation for some (or even all) of the observed astronomical properties of the universe that does not depend upon knowing the initial state of the universe (or whether it had an initial state), then, conversely, those same observations will be unable to tell us about the structure (or existence) of the initial state. We can't have our cake and eat it.

The inflationary universe can be seen as the type of answer the 'chaotic cosmologists' were seeking—but with a subtle difference. The chaotic cosmologists were looking for a way of damping down irregularities by physical processes. The inflationary universe shows how it is possible for our entire visible universe to be the expanded image of a primordial region so small that physical processes will keep it smooth, apart from very small statistical fluctuations. Hence, the expanded image of that tiny region displays the high degree of regularity that we observe, together with the small fluctuations. No irregularities are damped out by friction. If they existed before inflation occurred, they are still there; but they have been pushed beyond our visible horizon. We cannot see them.

In this way, inflation is able to provide an explanation for the gross properties of the visible universe, largely irrespective of how it began. So long as conditions arise which allow a small region to inflate for long enough, it will produce a large, smooth universe containing small irregularities, which expands very close to the critical divide between open and closed universes. From Fig. 6.8 we can see how the irregularity of the universe over very large scales is irrelevant once inflation begins. It takes a tiny piece of the Universe and accelerates its expansion so that it is bigger than our horizon today.

We can now understand why the inflationary universe places even greater limits upon our ability to determine the structure of the universe in the very distant past. We have already seen that the finite speed of light limits the region of space and time from which astronomers can obtain information. But inflation removes the information about the structure of the visible universe at times before inflation occurred.

Inflation is a very appealing idea if you are an astrophysicist seeking a simple explanation for how galaxies formed, or if you want to understand why the visible universe looks so similar in all directions. But if you want to know what the universe was like before inflation occurred (say, earlier than 10^{-35} seconds), or seek to determine whether the visible universe had a beginning, or to find

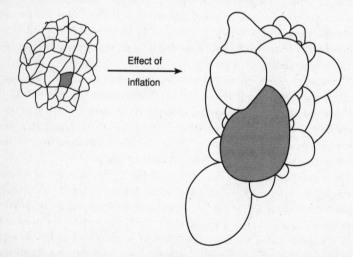

Fig. 6.9 The spatial structure of a chaotic inflationary universe. Different regions undergo different amounts of inflation. We are imagined to live within one of these large smooth inflated regions (shaded). Beyond our horizon other regions should exist with different densities and expansion rates, depending upon the amount of inflation they have experienced.

relics from the very early universe that shed light on the physics of elementary particles at energies greater than 10^{15} GeV, then the inflation is very bad news.

Thus, economics and cosmology conspire to place fundamental limits on our ability to probe the behaviour of matter at ultra-high energies. Faced with the huge costs of creating high energies on Earth, particle physicists have long hoped that cosmology would provide an inexpensive 'laboratory' in which to explore the structure of Theories of Everything. But, just as the smallness of our past light-cone stops us from drawing any conclusions about the structure or the origins of the entire Universe, so inflation will wipe the visible universe clean of the information we need to elucidate the ultimate laws of high-energy physics.

Inflation acts as a cosmological filter. It pushes information about the initial structure of the Universe out beyond our present horizon where we cannot see it; then, it overwrites the region that we can see with new information. It is the ultimate cosmic censor.

Chaotic inflation

> There was a young man of Cadiz
> Who inferred that life is what it is,
> For he early had learnt,
> If it were what it weren't,
> It could not be that which it is.

ANONYMOUS

We have highlighted the restriction of the scientific enterprise to the study of the visible part of the Universe within our horizon. But is this restriction worth worrying about? How much information could be lost to us because of this restriction imposed by the finiteness of the speed of light?

Before the possibility of inflation was discovered, it was generally assumed that the Universe should look pretty much the same beyond our horizon as it does inside it. To assume otherwise would have been tantamount to assuming that we occupied a special place in the Universe—a temptation that Copernicus taught us to resist. While it was grudgingly admitted that we might be wrong about this, such a view was regarded as rather pedantically positivistic. This philosophical attitude has been transformed. The general character of inflationary universes reveals that we must expect the Universe to be far more exotically structured in both space and time than we had previously expected.

If the Universe began in a chaotically irregular state, then some regions would undergo inflation, while some might not. The amount of inflation would vary from region to region, and the result would be a post-inflationary universe that was very different from place to place. Each inflated region would be like a bubble in which conditions would be smooth (the more the inflation, the smoother it would become) but different from those to be found in other bubbles. Our bubble must be very large, bigger than our horizon, but beyond it there should be other bubbles of different sizes, in which conditions differ from those within our own. For a schematic picture, see Fig. 6.9.

As the implications of this scenario have been explored, it has emerged that all sorts of other properties of the Universe could vary from one inflated bubble to the next. Some of the quantities that we call the 'constants of physics'—the strength of gravity, the masses of elementary particles, or even the number of dimensions of space—could vary from bubble to bubble.[16] All the astronomical observations that we have made show the values of constants of Nature to be the same from place to place within our visible horizon to huge precision (better than one part in 10^{15} in some cases). This is exactly what we would expect to find, even if those constants could vary all over the Universe. When any tiny region inflated, then all observers who eventually evolved within it would find the constants of Nature to be the same to very high precision because they derived from the same inflationary patch long ago.

This picture greatly extends our picture of the possible spatial complexity of the Universe. Unfortunately, it puts that complexity beyond the reach of science. One day, astronomers in the far future may see the signs of the nearest bubble coming into view. But they will never know how much more lies beyond.

Is the Universe open or closed?

> *I don't pretend to understand the Universe—it's a great deal bigger than I am.*
> THOMAS CARLYLE[17]

One of the starkest questions posed by the Big Bang picture of the expanding universe is whether our universe is going to continue expanding for ever or whether it is doomed to collapse back upon itself towards a 'Big Crunch' at some time in the future. These two alternatives are separated by the 'critical' universe. A critical universe possesses an exact balance between expansion energy and the gravitational pull of the matter within it. In 'open' universes the expansion energy overcomes the gravitational pull, while in 'closed' universes it is gravity that prevails. By making measurements of the expansion rate of the universe, and counting up all the matter our telescopes can detect, we might hope to decide which is the winner: expansion or gravity. Unfortunately, this is not so easy. Astronomy is about detecting light, but most of the matter in the universe appears to be dark. There is far too little luminous material to close the visible universe, but there might well be enough dark material hiding between the galaxies.

We have already seen that the universe is expanding close to the critical divide separating open and closed universes. So far, our observations are not accurate enough to choose one way or the other. If however, the inflationary universe theory is correct we shall *never* know whether the universe will expand for ever or contract.

Inflation predicts that any large region of the Universe, large enough to contain our visible universe, should now be expanding at a rate that is within one part in 100,000 of the critical rate. (It does not, however, say on which side of the divide we lie; that was fixed at the beginning of the Universe and cannot be altered.) No foreseeable astronomical observations are going to be accurate enough to discover on which side we lie. However, even if they were, it would not answer the question for us. The difference between the density of the visible universe and the critical value that is predicted in inflationary universe models is of the same magnitude as (or smaller than) the variations in the density that inflation produces from place to place. We expect variations in the density over the entire Universe beyond our horizon to be at least as great. This means that if, with perfect instruments, we could audit all the matter in the visible universe today and found it to be less that the critical density by one part in 100,000, this would not mean that the Universe is open and going to expand for ever. Any such conclusion would assume that the Universe beyond our horizon is identical to that within it. If we carried out the same audit of the mass one day later, when our visible universe was one light day larger in size,[18] then we might well find that the new material that had come within view was enough to increase the observed density to one part in 100,000 *above* the critical density, so making the Universe look closed, and destined to implode in the future. But, again, no such conclusion about the whole Universe would be justified.

The scales are very finely balanced. Our visible universe expands close to the critical divide; small fluctuations in density can decide the overall balance we discern between expansion energy and gravity. The visible universe could be an under-dense, open 'bubble' in an over-dense closed Universe; equally, it could be a closed bubble inside an open Universe. Observational astronomy can never tell us whether the entire Universe is going to expand for ever or whether it is finite or infinite. Even if we were to hit the final Big Crunch, we would not know how much of the rest of the Universe was sharing that fate.

Eternal inflation

The poet only asks to get his head into the heavens. It is the logician who seeks to get the heavens into his head. And it is his head that splits.

G.K. CHESTERTON

The complicated spatial variations that chaotic inflation would be expected to spawn in the early Universe are not yet the end of the story. Andrei Linde has discovered that inflation has a tendency to be self-reproducing.[19] Remarkably, it

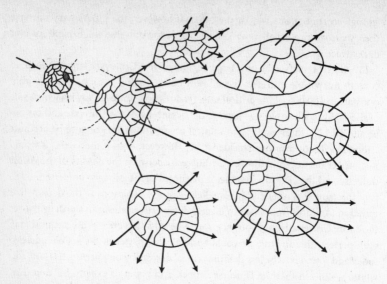

Fig. 6.10 A schematic representation of an eternal, self-reproducing inflationary universe. Each region that inflates naturally creates within it the conditions that guarantee that subregions of it will themselves undergo further inflation, and so on *ad infinitum*.

appears that the fluctuations that inflation produces have a form that inevitably induces further inflation to occur from small subregions of the bubbles that are already inflating. Inflation appears to be a potentially unending, self-reproducing process: in short, an epidemic (Fig. 6.10). Each bubble that is produced somewhere in space and time during this process can possess different values of many of its constants of Nature, defining the form of the physical structures that can arise within it. The Universe thus appears likely to be far, far more complicated in its historical development, as well as in its spatial variation, than we had suspected.

So far, our mathematical investigations show that this multiplication process will be endless, although individual bubble 'universes' may collapse to destruction if they are dense enough. Nevertheless, if we try to reconstruct the past history of this exotic evolutionary process, then events are not so clear. We have been unable to discover whether it must have a beginning in time. Probably, the whole self-reproducing network of inflating bubble universes need have no beginning, but particular bubbles may have beginnings when their histories are traced backwards. These beginnings would correspond to quantum-mechanical fluctuations in the energy of the Universe from place to place, and would appear spontaneously, with some probability, from time to time.

Our own place in this fantastic eternal panorama is an intriguing one. We are living in a particular bubble, with a particular suite of physical constants, at a particular epoch in this seemingly eternal sequence of inflations. We know only that we must be residing in a bubble that inflated long enough, and grew large enough, for stars to produce the elements upon which all known forms of complexity and life are based. We have no real idea how 'likely' this type of large bubble is, nor how probable it is that the constants of physics within it should fall out in a bio-friendly fashion.[20] The irony is that we shall never know whether or not this self-reproducing Universe, in all its baroque complexity, really exists or not. Our view of the Universe is confined by the finiteness of the speed of light and is doomed to be a parochial one.

Frustratingly, we now have positive reasons to expect the Universe to be extremely complicated in structure on all scales and at all times, and for our visible part of it to be atypical in important respects. The limits imposed by the finiteness of the speed of light prevent us from ever testing our expectations about the structure of the Universe beyond our horizon. As a result, the eternal inflationary universe picture cannot be tested by observations in the way that we can test whether we are living in a bubble that underwent inflation in the past—by looking at the detailed signature of the small variations in the temperature of the cosmic radiation around us. It is destined to remain an inspiring story.[21]

If the satellites scheduled to fly during the next ten years can prove conclusively that our visible universe does not bear the hallmarks of past inflation, then eternal inflation will lose its credibility. We could be living in a strange bubble that did not inflate, but that *ad hoc* possibility would not be enough to sustain our belief in a world of worlds beyond the horizon. But, if those satellites confirm our own inflationary past, they will strengthen speculation about the unobservable beyond, even though they will be unable to provide any data about it.[22]

Despite these limitations on our cosmological knowledge, the importance of eternal inflation, as a plausible cosmic scenario, is that it shows that there is every reason to believe that the restriction of astronomical observation to data within our horizon is a dramatic restriction upon our knowledge of the overall structure of the Universe. The Universe should be very different beyond our horizon. Ultimate questions about its beginning and its end are therefore doubly impossible to answer.

The eternal inflationary universe theory is often compared with the long-defunct steady-state theory of the universe, first proposed in 1948 by Herman Bondi, Thomas Gold, and Fred Hoyle.[23] In this theory of the expanding Universe, there was no Big Bang: matter was supposed to be continuously created at a rate that was sufficient to keep the density of the Universe constant at all times. The required creation rate is actually very small, about one atom in every cubic metre of space every ten billion years—far lower than could be detected directly. There was no beginning, and no end, to the steady-state universe. On the average it should appear the same to all observers at all times and in all places. It was like an inflationary universe that was always inflating. The standard Big Bang universe was a total contrast: all the matter came into being at some initial time and then expanded, cooled, and rarefied. Unlike the steady-state model, the past differed from the future. In the past the Big Bang universe was hotter, denser, and uninhabitable by creatures like ourselves.

The Bondi, Gold, and Hoyle version of the steady-state theory was eventually ruled out by many different observations. Populations of cosmic objects, like radio galaxies and quasars, were found to have birth rates that changed with cosmic epoch. But, most dramatically, the discovery of the microwave background radiation in 1965 demonstrated that the Universe had a past that was hotter and denser than the present.

As support for the steady-state theory withered away in the face of these new observational facts, some of its defenders became desperate to save it. One proposal was that, while the *observable part of the Universe* is of Big Bang type, it is merely an expanding Big Bang bubble within an infinite steady-state Universe.[24] No observations could confirm or refute this idea if the scale of steadiness was large enough.

Sometimes this old idea is compared with the chaotic or eternal inflationary theories. It is important to remember that there are important differences. The steady-state bubble universe was conceived purely at a last-ditch effort to save a cherished idea from being ruled out by observations. There were no scientific reasons for advocating Big Bang bubbles in a steady-state environment. Indeed, it was a step that went against the whole spirit of the steady-state philosophy, which hypothesized a Universe that was always and everywhere the same on the average.[25] By contrast, the extension of the inflationary universe to its chaotic and eternal forms was not proposed to rescue simpler versions from adverse observational facts. It emerged as an inevitable (and, to many minds, unwelcome) logical consequence of a theory that was not beset by observational problems.

The natural selection of universes

Anything that is produced by evolution is bound to be a bit of a mess.
SYDNEY BRENNER[26]

The self-reproducing inflationary universe introduces a wider consideration into cosmology. The sciences fall into two categories, split by their attitude towards explaining the existence of complexity. There are those, like biology, where the explanation for complex structures among living things is happily attributed to the process of natural selection acting over very long timescales. Very small variations occur, perhaps randomly, at each stage and the advantageous ones are passed on with a higher probability than the less advantageous ones. There is some serious current debate over the role that the self-organizing character of physical processes (which we looked at in the last chapter for the case of the sand pile) might also play in this process, but most evolutionary biologists seem strongly resistant to this possibility.[27] By contrast, the explanation of complex structures by evolution and natural selection has played no role in astronomy. The structure of objects like stars and galaxies is primarily governed by the laws of physics. They are equilibrium states between opposing forces of Nature. The important objects that physicists study, things like molecules, atoms, nucleons, and elementary particles, have invariant properties defined by the values of the constants and the laws of Nature. They do not possess the capacity for variation possessed by genes. Physicists and astronomers have therefore been led to expect that anything truly fundamental must be explained by some direct feature of the laws governing the four forces of Nature. They would be disappointed to find that some fundamental aspect of the visible universe had a messy explanation as the leftovers of some selection process. Biology, by contrast, is all about this messy business.

We have already discussed (in the previous chapter) some of the speculative ideas of Smolin and Harrison about the ways in which the Universe, and its defining constants, might evolve by some form of selection process. In Smolin's case, the selection might be called 'natural', with more black holes increasing the chance of survival of a particular variety of universe; but, in Harrison's case changes are brought about by the conscious intervention of intelligent minds, and might be more appropriately termed the 'artificial selection', or 'forced breeding', of bubble universes.

Linde's self-reproducing inflation can also be cast in the guise of evolution by natural selection.[28] Each inflating bubble gives rise to progeny which themselves inflate (*reproduction*). These baby universes possess small variations in the values of their defining constants of physics and other properties (*variation*), but they also carry some memory of the defining constants of physics in the bubble that gave birth to them (*inheritance*). The varieties of bubble which produce the most baby universes will be the ones that multiply and dominate the Universe in the long run. Strangest of all, if we believe the biologists who tell us that the definition of life is a process that possesses *reproduction*, *variation*, and *inheritance*,[29] then the self-reproducing inflationary universe is alive!

Topology

A burleycue dancer, a pip
Named Virginia, could peel in a zip;
  But she read science fiction
  And died of constriction
Attempting a Möbius strip.

CYRIL KORNBLUTH

We are used to spaces being curved in some way. We know that the Earth's surface is curved. The page of this book is curved. And, in the early years of this century, Einstein taught us that the whole Universe of space and time is curved by the presence of matter within it. What does this mean? We might gain an image of curved space by thinking of it as resembling a rubber sheet that is deformed when a mass is placed upon it. All masses have a purely local effect upon the geometry of space. This effect is manifested when light or other masses pass by. They keep sensing the shortest route to take, just like the stream flowing down the hillside, but the presence of the curvature, contributed by the mass, creates a shortest path that appears to be the result of an attractive force ('gravity').

Before Einstein's radical proposal, we thought that space was an unchanging stage upon which all the motions and interactions of matter were played out. Space was a tabletop rather than a deformable rubber sheet. And there is a real difference. Spin a ball on the tabletop and it will not affect another ball at rest somewhere else on the table. But spin a ball on the rubber sheet and its spin will twist the sheet, and cause nearby objects to be twisted round in the direction of the ball's spin.

Curved space is understandable by means of these analogies; but 'curved time' sounds odd. It means that the rate of flow of time is determined by the strength of the gravitational field where it is being measured. Time passes more slowly in strong gravity fields (where space is greatly curved) compared with its rate of flow in regions with weak gravity fields (where space is almost perfectly flat).

The curvature of space and time is a matter of geometry. Einstein's equations tell us how to calculate the curvature of space and time from any distribution of matter and energy that we care to prescribe (in principle!—in practice, the equations are very difficult to solve and we have been able to do this calculation only for very simple distributions of matter with a high degree of symmetry). The simplest models for the universe (for instance, the one expanding at the critical rate everywhere) describe an expanding space (the rubber sheet is being stretched) that looks like a completely flat, unbounded sheet at any instant of time. Now the local geometry of this space would be unaffected if the sheet were rolled to form a cylinder. Locally it would still appear flat: the three interior angles of any triangle would always add up to 180 degrees. But something has changed.

There has been a change in the *topology* of the Universe. Topology is changed only by tearing, cutting holes, or gluing parts of space together. If two surfaces can be deformed into one another by stretching, without tearing, then they possess the same topology. Hence, a ring doughnut is topologically equivalent to a coffee cup, but neither is equivalent to a cup with two handles. Figure 6.11 shows what occurs if we join two sides of the space. We could also join the other two perpendicular directions as well. The result is sometimes called the 3-torus topology ('torus' is just the mathematician's name for a ring doughnut shape).

We would like to know what the topology of our Universe is, but Einstein's equations are silent about this aspect. They tell us how to determine the geometry of the Universe from the distribution of stars and galaxies, but that does not tell us the topology. Astronomers generally assume, for simplicity, that if the Universe is critical or open then the topology is just that of an infinite flat sheet, and they call this the 'natural' topology. However, although this makes life easy for astronomers, there is no reason why space needs to be like this. If it were joined up like a cylinder in all three directions, it would have a *finite* volume even though its expansion would behave like that of a critical or open universe.

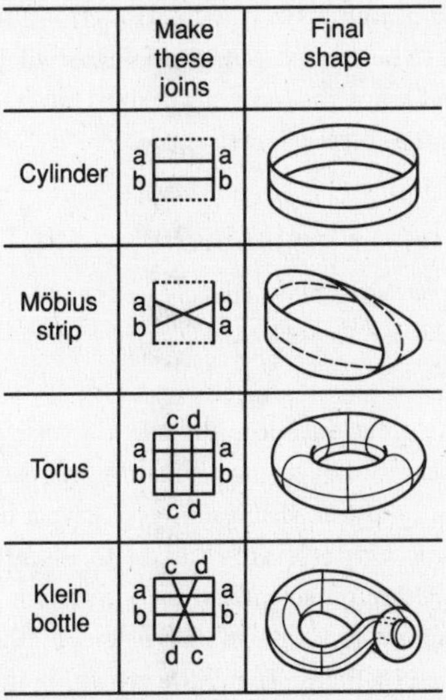

Fig. 6.11 Some spaces with unusual topologies. In each case a flat surface has two or four of its faces joined, either with or without a twist, to create four different topologies.

Since there are so many more ways for the topology to be unnatural than for it to be natural, we might even regard it as more likely to be unnatural.[30] Astronomy can help us, though; we can look and see whether there is any observational evidence for a 3-torus topology. Such a topology would lead to multiple images of distant galaxies, quasars, and clusters, as the light winds round and round the cylinder. It is like looking at yourself when standing between two parallel mirrors: you see an unending sequence of images getting smaller and smaller. So far there is no evidence of any repeated images that could be attributed to this phenomenon. Recently, some of us examined what would happen to the background radiation from the early stages of the expansion if the topology were unnatural.[31] We discovered that the maps of the sky that the *COBE* satellite measured would have a completely different structure if the topology of our Universe deviated from that of a flat sheet over a dimension smaller than about fifteen billion light years. If the Universe does possess an unusual topology, then its identifying features appear to be hidden beyond our visible horizon today.

The information needed to determine the overall topology of the Universe is inaccessible to us. We can place limits on the scale over which it can exhibit handles, or links back on itself, but we can never make the observations needed to determine its overall character. This is unfortunate because the fascinating question of whether physics can provide a description of how or why a Universe can be created out of 'nothing' is likely to be strongly influenced by the topology of the Universe that is to be created. Some topologies will be more likely to arise than others. If we don't know the topology of our Universe we might be lacking an important piece in the cosmic jigsaw puzzle.

Did the Universe have a beginning?

In the beginning there was nothing. And the Lord said: 'Let there be light' and there was still nothing, but now you could see it.

TERRY PRATCHETT

We have seen that the great question 'did the Universe have a beginning?' is unanswerable by observational science; but this still leaves us to ponder the more modest question 'did our visible universe have a beginning?' Questions about the origin of the Universe are difficult to disentangle from inherited religious prejudices. Although scientists do not set out to confirm or refute religious or mythological accounts of the origin of the Universe, they are undoubtedly influenced by them. They have grown up immersed in particular modern cultures, and become familiar with traditional speculations and dogmas. The stories they contain tend to suggest directions in which cosmological theories can be developed.[32]

Western cultures have several religious traditions in which the world has a beginning. For a thousand years, theologians have argued about the interpretations of these accounts and about the meaning of a 'beginning' for the Universe, together with such subtleties as whether time was created with the Universe or not. As a result, the idea of the creation of the Universe out of nothing is a familiar one, with which many people—scientists and non-scientists—feel comfortable. This does not mean that it is understood or logically self-consistent (the idea of unicorns does not make me feel uncomfortable either): simply that the general idea seems to be a cogent one.

This cultural background provided a fertile environment for the picture of an expanding Universe. It naturally supports the idea of a universe that 'began' a finite time ago. If we had found ourselves in a universe that was static, then we would have found it harder to reconcile it with our inherited beliefs about the Universe having a beginning.

Our inherited religious beliefs (or antipathy to them) can make us more inclined to develop modern mathematical cosmology in certain directions. Some cosmologists look for models with a beginning and seek a mathematical characterization of the beginning; others regard the prediction of a beginning as a sign that the theory is breaking down under extreme conditions, and seek to avoid it by changing the theory of gravity in some new way. For them, a modified theory, which does away with the singularity at the beginning of time, is an improved theory. By contrast, there are physicists, like Roger Penrose, who regard the singular beginning of the Universe as an important ingredient in its structure, without which we would miss some of its defining characteristics.[33]

In the period between 1922 and 1965, there was considerable confusion about the interpretation of the singular beginning to the Universe that is implied by the simple expanding universe models (shown in Fig. 6.2). They all expand, and, if we run the expansion backwards in time, we discover a state of infinite density and zero size at a finite time in our past.

The scientific basis for a beginning was not in itself new. In the nineteenth century the early investigators of thermodynamics had applied the second law of thermodynamics to deduce that there must have been a past moment of maximum order, which they interpreted as a beginning.[34] In fact, this argument is not quite right. The entropy does not need to have a minimum just because it is always increasing.[35]

At first, many cosmologists suspected that the 'beginning' predicted by the expanding universe models was not real.[36] Three objections were raised. Some argued that the inclusion of cosmic material with realistic pressure would resist compression to zero size (just like squeezing a balloon) and the Universe would have 'bounced' at a finite radius. When universes with pressure were investigated, it was found that they had a beginning of zero size as well. Ordinary pressure didn't help to evade the singularity.

Next, it was argued that the beginning was an artefact of considering universe models which expanded at the same rate in every direction. When you ran them backwards everything piled up at the same point. Again, other universe models were investigated. Some expanded at different rates in different directions, others varied from place to place as well. In all cases, the singular beginning remained.

Finally, a more subtle objection was raised. Suppose the singularity of zero size was just a breakdown in our way of mapping and describing the Universe, rather than in any physical feature of the Universe itself. A similar dichotomy arises when we look at the lines of latitude and longitude that we employ to map positions on the surface of the Earth. As we move towards the poles on a geographers' globe, the meridians approach one another, and finally intersect. The map coordinates are singular at the two Poles. But this does not mean that anything strange happens to the Earth's surface there. Polar explorers can change to a new system of map coordinates more convenient for their purposes if they wish. How can we tell that the Big Bang singularity is not also of this innocuous variety?

The answer was given by Roger Penrose, who found a new way to attack the problem which sidestepped the need to worry about coordinates and asymmetries in the universe.[37] With Stephen Hawking, he showed that if a number of reasonable assumptions about the universe hold, then there must be a beginning.[38] The most important of the reasonable assumptions was that gravity must always be attractive. If gravity is attractive, time travel is impossible, and if there is enough matter and radiation in the universe today, then there must be at least one path through space and time taken by light rays or massive particles that cannot be extended indefinitely into the past: it must have a beginning.

Strictly, only one historical path need have a beginning, and no extremes of temperature and density need accompany it. The theorem cannot tell us things like that. In practice, cosmologists believe that these extreme physical conditions would accompany a beginning, if there was one. The conventional Big Bang picture of a universal beginning which is physically extreme everywhere is completely consistent with the theorem of Hawking and Penrose, although not strictly demanded by it. The logic of the theorem is important. It is a theorem not a theory. If its assumptions hold, then there *must* be a finite past history. If the assumptions do not hold (for example, if there is some form of matter in the universe that is not gravitationally attractive), then we can conclude nothing at all: there may be a beginning, but then again there may not.

In the period from 1966 to 1975, this 'singularity theorem' provided very strong grounds for believing that our visible universe (which most commentators did not then distinguish from the whole Universe) had a beginning. The key assumptions of the theorem—that gravity was attractive, and that the universe contains enough matter—seemed to be true. The microwave background radiation turned out to provide enough mass-energy, and all forms of matter that physicists had encountered experimentally, or conceived of theoretically, exhibited gravitational attraction. The theorem seemed to apply to our universe.

Then, after 1975, things began to change. Cosmologists began to think seriously about how the effects of quantum uncertainty might affect the attractive nature of gravity. New investigations into elementary particle physics at very high energies inspired new attempts to reconstruct the very early history of the universe, and to find ways of using astronomical observations to test those theories. These investigations led to Alan Guth's proposal of the inflationary universe which we have already met.[39] One thing became abundantly clear from these studies: our theories of elementary particles lead inevitably to the existence of new types of matter in Nature. These new particles, the scalar fields that drive inflation, could display negative pressure, and if they changed very slowly in the early universe they would antigravitate. As a result, they violated the assumptions of the Hawking–Penrose theorem. Moreover, it was precisely this possibility that allowed them to accelerate the expansion of the early universe for a brief period and thus to create the phenomenon of inflation. Suddenly, cosmologists no longer believed that the key assumption of the Hawking–Penrose theorem would hold in Nature. Particle physics theories provided many plausible matter fields which would antigravitate at very high energies and this antigravitation would allow all the benefits of a period of inflation to be reaped.

The simple conclusion was that, if we wanted to have inflation, we could not draw any conclusions about a singularity at earlier times. The violation of the requirement that all matter fields are gravitationally attractive does not mean that there was no past singularity, only that we can no longer tell one way or the other. In fact, we have seen how eternal inflation leads to the complicated picture of the Universe as a 'multiverse' budding small 'baby' universes, some of which inflate to become large like our own visible universe, while others just collapse and dissolve into a foam of space and time. This process appears to have no end, but did it have a beginning? The answer is not yet clear.

No one doubts that these extrapolations backwards in time will eventually break down because our knowledge of high-energy physics is incomplete, or untestable, in some way. From what we discuss in this and the last chapter, there are good reasons to expect that we shall remain ignorant of some of the things that we need to know in order to determine whether our visible part of the Universe had a beginning or not.

One of the interesting features of Einstein's theory of gravitation is that it predicts that it cannot predict. There is a time in the past before which the assumptions upon which Einstein's theory of gravity is based must fail. Space and time become subject to quantum uncertainties which are ignored in Einstein's theory whenever it is used to examine the structure of space over distances smaller than 10^{-33} cm, intervals of time shorter than 10^{-43} seconds, or energies exceeding 10^{19} GeV. These frontiers are called the Planck scales, after Max Planck, the great German physicist who pioneered the development of quantum theory. When we get within 10^{-43} seconds of the apparent beginning of the universe, Einstein's theory of gravity fails. We don't know whether the singularity it allows is a consequence of the theory breaking down or whether it is a real physical occurrence. To probe further back in time, a quantum theory of gravity is needed. Currently, there are different approaches to this issue. Superstring theory offers the most attractive route, but it is not known whether it permits singular universes or not.

Naked singularities: the final frontier

The amount of eccentricity in a society has been proportional to the amount of genius, material vigour and moral courage which it contains.

JOHN LOCKE

The past of the Universe is not the only place to look for singularities. Every time a star with a mass more than about three times that of the Sun exhausts its nuclear fuel supplies and begins to contract under its own gravity, a singularity may form. In fact, this was the situation to which Roger Penrose's first 'singularity theorem' applied. At first, one might think that this opens up the possibility of observing what is going on very close to a singularity and using this information to understand what might have occurred near a cosmological singularity in the past.

Alas, this appears to be impossible. When massive aggregates of matter collapse under gravity they eventually compress such a large amount of mass into so small a region of space that nothing can overcome the pull of gravity and escape—not even light. There is a surface of no return—called an 'event horizon'—beyond which nothing can be recovered. The region within the event horizon is called a *black hole*. Astronomers believe that several have been identified.[40] When they orbit around ordinary stars they pull material away from the surface of their companion star in distinctive ways and create rapidly flickering X-rays that reveal their characteristic size and gravitational presence.

Once the event horizon of the black hole forms, outsiders see it as an unchanging source of gravitational pull. But our mathematics tells us that inside the horizon material will just keep falling into the centre, where the density will get higher and higher. Ultimately, our equations predict the occurrence of a singularity of infinite density, where space and time cease to exist, unless new laws of physics come into play in which gravity mingles with quantum uncertainty, just as at the beginning of the expansion of the Universe. Falling into the centre of a black hole is just like approaching the final Big Crunch of a closed universe.

This state of affairs illustrates a strange feature of the Universe. It appears to allow singularities to develop inside black holes formed from collapsing massive stars, but it surrounds them by event horizons which prevent the singularity from influencing the outside universe in any way. At first, this seems like an annoying limit on our ability to discover what happens near a singularity. But, upon reflection, it may be necessary for the rational self-consistency of the Universe. Singularities are, by definition, places where the laws of physics break down. Anything can come out of a singularity—TV sets, time machines, whole universes even—there are no known rules. If such a singularity were present near by, we would be unable to use the laws of Nature to predict the future. Black holes are our defence. We are protected from the totally unpredictable effects of local singularities, which may form every time a massive star collapses in our Galaxy, by the formation of event horizons. Science fiction films always dramatize the event horizon as a sort of cosmic Venus flytrap. Its real importance is as a shield against what would otherwise come out into the Universe.

So important is the presence of an event horizon around a singularity that Roger Penrose has argued that singularities can never be 'naked', but must always be clothed by an event horizon. This veto on naked singularities is called the hypothesis of *cosmic censorship*. There exist proofs of various versions of it, but it is not known whether it is generally true, even when we ignore the influence of quantum physics on gravitation. When we include quantum physics it may not be true at all. Stephen Hawking showed that the inclusion of quantum processes into the study of black holes means that they are not really 'black'.[41] They radiate particles and radiation from their surfaces, and slowly evaporate away. As their mass falls, the temperature of the radiated particles, and the evaporation rate, increases. The event horizon gets smaller and smaller. When the Planck scales are reached, the black hole explodes like a miniature version of the Big Bang. What remains we do not know. But if one of these black hole explosions could be observed locally, it would provide a momentary glimpse of physics near the Planck scale, without the shield of an event horizon.

These black hole explosions cannot result from the evaporation of the black holes that form from the death of massive stars. They are too massive and their evaporation rate is far too slow. Rather, they would be the end point of the evaporation of much smaller black holes that could have formed only in the very early universe. Black holes close to about 10^{14} g in mass (about the mass of a large mountain), which could only have formed when the Universe was about 10^{-23} seconds old, would be in the final throes of explosive evaporation today. Astronomers have looked for these exploding black holes in many different ways. They would emit high-energy gamma rays and create bursts of strong radio waves and cosmic rays. So far, there is no positive evidence that any of these outbursts have been seen at the level that our telescopes can detect. All we can say is that, if they do exist, there must be fewer than one explosion in every cubic parsec (more than 29×10^{39} km^3) of space per year. The theoretical expectations are not too hopeful either. If inflation occurred in the very early Universe it would have smoothed out irregularities to such an extent over regions encompassing about 10^{14} g in mass that they would have been unable to collapse in upon themselves to create these small black holes. But some maverick versions of inflation predict that these very small black holes might form at the end of a period of inflation.

Dimensions

The first thing to realise about parallel universes ... is that they are not parallel. It is also important to realise that they are not, strictly speaking, universes either, but it is easiest if you try and realise that a little later, after you've realised that everything you've realised up to that moment is not true.

DOUGLAS ADAMS[42]

The problem of discovering the structure of our visible universe near the Planck scale is like unwrapping a sequence of Russian dolls. As we work backwards, we keep encountering new limits that prevent us from delving further backwards. At first, the limits are mainly inconveniences. The opaqueness of the universe to photons means that we have to look for the products of nucleosynthesis. But inflation erects a serious shield. If inflation turns out to be one of those beautifully simple ideas that the Architect of the Universe chose not to include in his plans, then we might be able to look farther back by observing gravitons flying freely to us through space and time from the Planck scale. But superstring theory has opened another Pandora's box of possibilities. Superstring theories are the only current theories of physics which do not lead to internal contradictions or to predictions that measurable quantities have infinite values when gravity is merged with the other forces of Nature. Yet these consistent theories of the fundamental forces of Nature appear to require the Universe to have many more dimensions of space than the three that we habitually experience. The original string theories required the Universe to have either 9 or 25 dimensions of space! Since we see only three dimensions we must either conclude that these theories are wrong, that dimensions can be something other than what we are used to thinking them to be, or that lots of dimensions of space are hiding somewhere. While either of the first two options might turn out to be the case, it is generally assumed that the third provides the answer to the conundrum. Some process must be found which allows three (and only three) of the total number of dimensions of space to grow very large while the rest remain trapped at the Planck scale of size, where their effects are imperceptible to us. In fact, when one looks more closely, it turns out to be more probable that *all* the dimensions should stay trapped at the Planck size. The conundrum is how three of them have become so much bigger: 10^{60} times bigger than the Planck size, in fact. What is required is a process which leads to the inflation of only three of the dimensions. At present no such selective process is known. This process might be random in character, so that its choice of three large dimensions was not programmed into the laws of physics. Alternatively, there might be a deep reason why three and only three dimensions can inflate. Inflationary universes can be concocted in which inflation occurs in different numbers of dimensions at different places, but they are rather artificial and unconvincing so far.

Putting the mystery of the selective inflation process to one side, we see that we are confronted with a major uncertainty. The true constants of Nature, and the forms of the laws of Nature, are really framed in 9 or 25, or some other number, of dimensions of space. A complicated physical process leaves only three expanding to constitute the astronomical universe we see around us. The quantities that we call the constants of physics are just three-dimensional shadows thrown by the true constants, which live in the full number of dimensions. Remarkably, if the extra dimensions exist and change their size by expanding as our three-dimensional part of the Universe does, then this would be revealed by a change in our 'constants' of Nature at exactly the same rate.[43]

The possibility that our Universe contains many more than three dimensions of space, trapped at the Planck scale of size, means that our access to the overall structure of the Universe might be limited even more dramatically than we have previously suspected.

Symmetry-breaking

The exact sciences start from the assumption that in the end it will always be possible to understand nature, even in every new field of experience, but that we may make no a priori assumption as to the meaning of 'understand'.

WERNER HEISENBERG

When we raised the possibility that the number of dimensions of space that grow large might be determined randomly, we entered into another feature of the way in which the Universe assumes its observed properties. Some of those present properties are reflections of the laws of physics and the properties of the most elementary particles of matter. Others may be the result of historical accidents which could have fallen out in some other way. These accidental outcomes can affect fundamental aspects of Nature. For example, the observed imbalance between matter and antimatter in the visible universe may be a direct reflection of an imbalance between matter and antimatter in the laws of Nature. Alternatively, part, or all, of the observed imbalance may arise from a random process. If so, then the imbalance would vary from place to place in the universe, and could not be predicted in the same way as it could if it were a universal consequence of the laws of Nature.

We have mentioned the possibility that the dimensions of space and the matter–antimatter balance in space might be historical accidents. It is also possible that the values of the constants of physics that we measure are accidents of the same sort. They might differ from place to place in the universe. At first, this seems implausible because we can check whether some of the constants of Nature could have been significantly different in the past by direct measurement. The strength of the electromagnetic force of Nature, and with it the whole of atomic and molecular structure, chemistry, and materials science, is determined by a pure number called the *fine structure constant*. It is equal to $(7.29735 \pm 0.00003) \times 10^{-3}$, or roughly 1/137. This is one of the famous unexplained numbers that characterize the universe. (I would bet that it occurs in the security codes, passwords, and pin numbers used by a surprisingly large number of physicists.)

We can measure the fine structure constant with very great precision, but so far none of our theories has provided an explanation of its measured value. One of the aims of superstring theory is to predict this quantity precisely. Any theory that could do that would be taken very seriously indeed as a potential 'Theory of Everything'.

Is the fine structure constant really constant? Laboratory experiments have shown that, if it is changing, then the rate of change relative to its present value is less than 3.7×10^{-14} per year.[44] Since the expanding universe is about 10^{10} years old, this means that it can have changed by less than one part in 10,000 over the whole of that time. Astronomy allows us to do even better. If we look out into the astronomical universe, we can observe quasars which first sent their light to us when the universe was billions of years younger, and several times smaller, than it is today. The detailed form of that light reveals the atomic properties in the intergalactic medium between the quasar and ourselves. The interrelationships are identical with those found by looking at the light emitted by the same pattern of elements in our laboratories down to the limiting of accuracy of the measurements.

Some of us have shown recently that if the fine structure constant is changing as the universe ages, then its must be changing relative to its present value at a rate less than 5×10^{-16} per year[44]. Moreover, by looking at different quasars in different directions in the sky we can also establish that the fine structure constant is the same from place to place over millions of parsecs of space to within on part in a million.

There is another way that we can get some information about the values of 'constants' of Nature in the past. Nearly two billion years ago, at the current site of an opencast uranium mine at Oklo in the West African Republic of Gabon, an accident of geology created conditions under which natural nuclear reactions occurred.[46] France had been mining uranium ore in their former French colony for many years when, in 1972, they extracted a sample which contained 71.71 per cent of the isotope uranium-235, instead of the expected 72.02 per cent. At first, theft or sabotage was suspected, but further investigation revealed that the isotope concentration had been depleted by the action of natural radioactive decay processes. In effect, a short-lived natural nuclear reactor was created by the peculiar geological conditions prevailing at the site. A

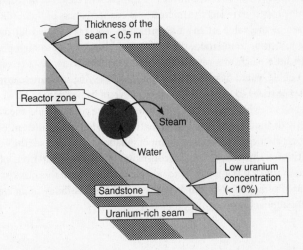

Fig. 6.12 The geological structure of the site of the Oklo natural reactor. A uranium-rich seam lies within a layer of sandstone lying on another layer of granite. The tilt of the granite and sandstone layers led to localized concentrations of uranium and water. When they exceed 10 per cent chain reactions can begin. The uranium layer must be thick enough to prevent the neutrons escaping, and be free from contamination by other heavy elements that would absorb all the neutrons. The nuclear reactions were self-regulated by the moderating presence of the water, which turned into steam when the reactions went faster, so slowing them down, but condensed when the reactions slowed down, so reducing the absorption of neutrons and speeding up the reactions gain.

very delicate arrangement of uranium ore embedded inside sandstone was lying on top of a granite that was dipping at 45 degrees (see Fig. 6.12). This tilt enabled uranium to accumulate to critical levels, whereupon chain reactions began, moderated by the presence of water.

The nuclear reactions that occurred left a signature of fallout that enables their reaction chain to be reconstructed. The key reaction is very remarkable. Its possibility hinges upon a coincidence that we know now exists between the different strengths of natural forces and the masses of nuclei. The fact that this unusual coincidence clearly also existed 1.8 billion years ago, when the nuclear reactions ran, enables us to set limits to the value of the fine structure constant at that time. We find that it could have differed from its present value by no more than one part in ten million; otherwise the natural reactor would not have functioned. If the fine structure constant is changing, then it must be changing at a rate that is less than 6.7×10^{-17} per year.[47]

This evidence is very persuasive. But does it imply that the fine structure constant is really constant all over the Universe? Unfortunately not; if the inflationary picture of the universal expansion is true, then the whole of our visible universe is the expanded image of a tiny causally coherent fluctuation. Its large-scale properties reflect the microscopic connectedness of that little patch. Thus, even if the value of the fine structure 'constant' varied all over the Universe just prior to inflation, the value would be constant to high precision over each little patch of Universe that underwent inflation. As a result, each inflated bubble would display the same value of the fine structure constant all over its space today to very high precision. Beyond its horizon, there would be other bubbles in which the value of the fine structure constant, like the value of the density or the level of density fluctuation, could be very different because they had experienced a different amount of inflation. Again, we see the very real possibility of a Universe in which 'constants' of Nature differ from place to place, but in which it is impossible to observe this because of the limits imposed by the speed of light and by our inability to test the possibility of an eternally chaotic inflationary process.

Summary

In one inconceivable complex cosmos, whenever a creature was faced with several possible courses of action, it took them all, thereby creating many distinct temporal dimensions and distinct histories of the cosmos. Since in every evolutionary sequence of the cosmos there were very many creatures, and each was constantly faced with many possible courses, and the combinations of all their courses were innumerable, an infinity of distinct universes exfoliated from every moment of every temporal sequence in this cosmos.

OLAF STAPLETON

In this chapter we have looked more deeply into the problems of cosmology that have challenged us for so long. Despite the success of Einstein's theory of gravity in describing the universe that we can see, we know that there exist fundamental limits to the cosmological quest. The finiteness of the speed of light partitions the Universe into parts which are out of causal contact with each other. We can gather information about the Universe only from the region within the horizon that the speed of light defines for us. This prevents us from ever answering deep questions about the origin or the global structure of the entire Universe. We cannot discover whether or not it is infinite, whether it had an origin in time, whether its entropy increases like that of small systems, or whether it is open or closed. Our observations are confined to determining the structure of the visible part of the Universe. Whereas this restriction would once have been regarded as unmotivated by what we know of the Universe, this is no longer the case. The inflationary universe theory, in all its developments, persuades us that we should expect to find the Universe complex in its spatial structure and in its temporal development. We appear likely to sit in a particular expanding bubble, unable to investigate the possibility that a Universe of elaborate never-ending complexity is blossoming beyond our horizon. Future satellite missions will provide decisive tests of the idea that we live within a bubble that has suffered inflation in the past, but we shall be unable to observe anything of other bubbles beyond our horizon. Finally, we have seen how the inflationary universe phenomenon, while providing explanations for several of the properties of the observable universe, prevents us from gathering information about events that preceded it. The origin of even our visible part of the Universe is hidden from us. We have found that the great theories of relativity and quantum mechanics combine to provide us with an account of the universe that we see, regardless of how it began. The price we have to pay for this unexpected gift is the relinquishing of information about how, or if, the Universe began and about all its properties beyond our horizon. The Universe is not only bigger than we can know, it is bigger than we can ever know.

CHAPTER 7

Deep limits

*In order to be able to draw a limit to thought, we should have to find both sides
of the limit thinkable . . . we would have to be able to think what cannot
be thought.*

LUDWIG WITTGENSTEIN

Patterns in reality

*He's the Master of Balliol College.
What he doesn't know just isn't knowledge.*

FELLOW OF ANOTHER COLLEGE

Any talk of limits to science will alarm many people and comfort others. There
are some who would equate the very idea of limits to scientific knowledge with a
violation of our freedom of thought and action. Limits of cost are one thing, but
absolute limits are surely something completely different. Show me one of those
and I'll jump over it, tunnel under it, or simply skirt round it. Yet, the more we
try to grasp what science is, and how it relates to the activity of human minds,
the more we are drawn towards the possibility that limits might be deeply
rooted in the nature of things. They might even *define* the nature of things.
Maybe this should not surprise us: science exists only because some things are
impossible.

Limits are slippery things. If we were to arrive at a definition of 'science' or of
'knowledge', then at once we would have circumscribed what we are going to
consider as scientifically knowable. We have established a limit. We might worry
that this dilemma afflicts any system of rules and regulations that we care to
follow. In some sense it does. If you set up a system of rules of reasoning, then
you are by definition placing limitations on what you will count as being true. It
should come as no surprise that there are limits to what can be deduced or
excluded by any system of rules. Keep only rabbits in your garden and you
should not be surprised if rabbits, and only rabbits, result.

Despite our grumblings, we like rules and regulations. Human cultures
abound with self-imposed constraints. We like to play games and solve puzzles;
we make music that is constrained by rigid rules of form. Art traditionally
explores the limits of a limited domain of space or time using prescribed

materials. Occasionally, an artist will invent a new arena, or ostentatiously break the bounds of the old idiom, but this is usually followed by a new exploration of the slightly enlarged domain with its own, albeit different, rules.

There is an interesting historical example of the awkward balancing act that is needed to think about limits of thought. Ancient philosophers and theologians used to struggle in their quest to talk about concepts like that of 'God', and there emerged a tradition of 'negative theology' which maintained that God transcended all descriptions. He was defined in terms of all the things that he was not: incomprehensible, atemporal, and so forth. One can see that this might be dangerous ground, for even to maintain that God is incomprehensible is to express a fact about God. To say that God is infinite seemed to be a way of ensuring that he possessed superhuman characteristics, but why can we not comprehend infinities? The natural numbers 1,2,3,4,5... are an unending infinite sequence, but this hardly renders them incomprehensible to us. Indeed, the entire discussion seems to be flawed by a belief that, to be valid, a description of something must share the quality it describes that, for example, an adequate account of the infinite must itself be infinite. However, hardly any of the descriptions we give of things around us have this unusually rigid property. The notion of being warm is not itself warm; the notion of being square is not square; and so on. Likewise, there is no immediate reason why the nature of unknowability should itself be unknowable.

Societies are regulated by laws and rules, and they take their character from the nature and number of those rules, the ways in which they are enforced, and the consequences of violating them. Most of us live in societies where everything that is not forbidden is allowed, rather than in tyrannical dictatorships where everything that is not demanded is forbidden.

The Universe also enjoys constraints. There appear to exist patterns of behaviour from which, in our experience, the Universe has never deviated. This should not surprise us. If there were no patterns of behaviour in the Universe then the total chaotic anarchy that would exist could not give rise to conscious intelligences like ourselves. No form of organized complexity could exist or develop unless there were limits to what can occur in the Universe.

The evolution of life by a slow process of natural selection is possible only because there exist regularities in Nature. Those regularities specify the reality to which the gradual process of adaptation approaches. Whether or not living things are aware of it, they are embodiments of theories about the laws of Nature drawn from the part of Nature that they have encountered. The size and strength of a bird's wing reflects the intrinsic strength of gravity about which the bird has no theoretical understanding. The structures of our eyes and ears embody truths about the phenomena that we call 'sound' and 'light', irrespective of our theories and beliefs about them.

Nature exhibits pattern, and so it obeys rules, and is subject to limits: there are things which cannot happen. Natural patterns allow us to build artificial patterns of still greater complexity. These patterns can be viewed as consequences of the natural pattern that we call 'life' interacting with other patterns. Our patterns of social behaviour are consequences of other, more loosely constrained, interactions between large numbers of examples of living complexity that is aware of itself.

The inevitability of pattern in any cognizable Universe means that there can exist descriptions of all these patterns. There can even be patterns in a collection of patterns, and patterns in the patterns in the collections of patterns, and so on. In order to describe these patterns, we need a catalogue of all possible patterns. And that catalogue we call *mathematics*. Its existence is not therefore a mystery: it is inevitable. In any universe in which order of any sort exists, and hence in any life-supporting Universe, there must be pattern, and so there must be mathematics.[1]

Some of the patterns that mathematics catalogues are shapes and symmetries that we can see, like in a tapestry or a mosaic. Others are relationships between more abstract entities: programs that relate lists of numbers to other lists, instructions which alter the shape of designs, logical relationships between properties of things, or patterns in sequences of quantities. Figure 7.1 illustrates some natural patterns.

This reveals why all discussions of the Universe and its contents lead so quickly and inevitably to mathematics: no science exists without it. That does not mean that all science has to be bristling with algebra and equations; sometimes the patterns that mathematics codifies can be handled quite easily with ordinary language. The words could be replaced by symbols and equality signs, but it is unnecessary. Only when the interrelationships become complicated, and the number of variables large, does it become expedient to replace them by symbols (just as ordinary language often resorts to using acronyms or abbreviations: OK?). If Nature reveals patterns, then there will be relationships between things, or habitual consequences of certain events occurring, and we can introduce symbols and rules to represent those relationships. Gradually, this process creates the structure which we call 'mathematics'.

Viewed as the gallery of all patterns, mathematics is something infinitely bigger than science. Science needs only some of the kaleidoscope of possible patterns to describe the physical universe. Thus, while mathematics can offer descriptions or predictions of the things that occur in Nature, it is not a science. It cannot tell us whether or not things exist in physical reality. This limitation shows up in the way mathematicians work. They are often happy to investigate structures which are unrealistic in the sense that they do not appear to offer a description or explanation of anything real. Remarkably, it has often turned out

Fig. 7.1 A collection of impressive natural patterns. (a) Crocodile; (b) honeycomb; (c) snowflake; (d) spider's web; (e) sunflower; (f) nautilus shell. (Photos: (a), (d), (e), (f) Planet Earth Pictures; (b) Tony Stone Images.)

that patterns which were originally investigated by mathematicians for purely aesthetic reasons turn out to play a key role in the structure of the physical world.

This leads us to think about two worlds: the physical world of matter and energy, and the mathematical 'world' we use to encode the patterns we find in the real world. Figure 7.2 is a schematic diagram of how the world of mathematics in which we represent patterns can be related to the physical world of events.[2] In each world there is a process of cause and effect; and ways of relating the assumptions and conclusions of the two realms. We try to encode physical events into the mathematical world, where deductions can be drawn, before decoding them back into the natural world.

Another curiosity of mathematics is that it is an unlimited human endeavour. We can imagine knowing all there is to know about geology, or the laws of physics, but it is difficult to imagine that mathematics could ever be closed. In fact, as we shall see, the unboundedness of mathematics is of a different and unexpected sort.

When we talk of understanding the world this means that we are able to replace a list of the facts by a pattern which links them all together in some way. It is inevitable that this pattern is part of the system that we call mathematics. As a result, any internal limitation that mathematics might possess can show up as a limitation on our ability to codify and understand patterns and their ramifications. Although we can distinguish the real world from the mathematical world, as in Fig. 7.2, as soon as we seek to understand the ordered world we are forced to establish points of contact between the two.

We shall see that the mathematical world contains all sorts of unexpected properties and restrictions on its full exploration. Parts of it cannot be pinned down and listed: they transcend the capabilities of any possible computer, and lie beyond the grasp of any quest to decide whether all its statements are true or false. We are going to explore some of the ramifications of such things for the

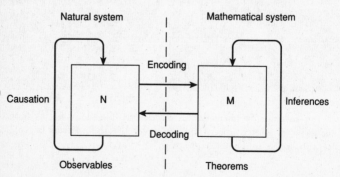

Fig. 7.2 A representation of the worlds of mathematics and Nature, which illustrates the process of mathematical modelling, devised by Robert Rosen[2].

research programmes of science. Whereas previous chapters have focused on the practical limits of science imposed by our own nature, and on the practical restrictions imposed by the place and time in the history of the Universe at which we find ourselves, now we are going to explore limits imposed by the nature of knowledge itself.

Paradoxes

> 'Is there a God, Lasher?'
> 'I do not know, Rowan. I have formed an opinion and it is yes,
> but it fills me with rage.'
> 'Why?'
> 'Because I am in pain, and if there is a God, he made this pain.'
> 'But he makes love, too, if he exists.'
> 'Yes. Love. Love is the source of my pain.'
>
> ANNE RICE[3]

Once we recognize that we have many possible logical systems at our disposal, we have to tread carefully. We can make statements in the language of one of our systems, but we are also able to make statements about that language using another one. For example, '2+2=4' is a statement *of* arithmetic; but, '2+2=5 is false' is a statement *about* arithmetic. Likewise, with human languages: we can talk in German about sentences in English. The language consisting of all the statements about another language is called its *metalanguage*. In our last example, German is being used as a metalanguage for English. Any metalanguage can in turn have its own metalanguage: I could write in Greek about someone else writing in German about sentences written in English. There is a never-ending hierarchy of metalanguages.

The distinction between languages and metalanguages is an important one if we are to identify the limits of the use of logic and what is meant by the concept of truth. Without this distinction, logic collapses into confusion, and any statement you care to make is 'true'. Suppose that you want to prove that the Earth is flat. Then just consider the following sentence:

Either this whole sentence is false or the Earth is flat.

This sentence is either true or false. If it is false then, by its own statement, the Earth must be flat. If it is true, then either the first statement 'this whole sentence is false' or the second statement 'the Earth is flat' must be true. Since we are now assuming the whole sentence to be true, the first possibility is excluded, and hence the second must be true. Therefore, the Earth is flat! Better still, you can replace 'the Earth is flat' with any other statement you care to choose and by the same reasoning, prove that it is true.

The Polish mathematician Alfred Tarski finally clarified this alarming situation in 1939. Statements in a particular logical language cannot be called true or false unless we step outside that language and use one of its meta-languages. If we want to say that a statement about the world is true, then we must use a metalanguage. Tarski proposed an unambiguous method of determining what we mean when we say that a statement is 'true'. He proposed that *the Earth is flat* is true if and only if the Earth is in fact flat. What this means is that the italicized sentence about the Earth is true if and only if the Earth can actually be demonstrated to be flat by replacing the word *Earth* in the sentence with the actual planet without changing its meaning. So, we can discuss the italicized sentence, debate about whether or not it is true, and test it against the geographical evidence, but the italicized sentence has no meaning until we do this in an unitalicized metalanguage.

This careful distinction removes all sorts of ancient linguistic paradoxes like 'this sentence is false'.[4] We see now that it is merely a confusion of a language with its metalanguage. The same flaw exists in our earlier example of '*Either this whole sentence is false or the Earth is flat*'. It mixes statements and (meta)statements about statements. The Earth is not flat after all.

This comforting conclusion has a more surprising by-product: there is no such thing as *Absolute Truth*. There are deductions that can be made within a language (proofs) which define what is meant by truth within that system, but there is no end to the hierarchy of metalanguages that tower above it, each with its own circumscribed area of truth. Tarski showed that it is impossible to construct a formal definition of truth or falsity. Truth cannot be rigorously defined in the same order of language that is used to express it, only in a metalanguage.[5]

These excursions seem far away from the world of science, but their influence was certainly felt there. The revelation that there were an infinite number of geometries and logics, all consistent but different, had a liberating effect on physics. The young Werner Heisenberg followed the development of thinking about alternative geometries and logics that undermined the idea of absolute truth. The possibility of alternative axiomatic bases for physics motivated Heisenberg in his search for a quantum-mechanical description of the world. Of their influences, he later wrote:

> I heard about the difficulties of the mathematicians. There it came up for the first time that one could have axioms for a logic that was different from classical logic and was still consistent . . . that was new to many people . . . I could not say there was a definite moment at which I realized that one needed a consistent scheme which, however, might be different from the axiomatics of Newtonian physics. It was not as simple as that. Only gradually, I think, in the minds of many physicists developed the idea that we can scarcely describe nature without having something

consistent, but we may be forced to describe nature by means of an axiomatic system which was thoroughly different from the old classical physics and even a logical system which was different from the old one.[6]

Heisenberg picks on the property of consistency as essential. We can easily create mathematical statements that are inconsistent ($0 = 1$), but what about physical inconsistency. What would it look like? Is it even conceivable?

Consistency

> *I think mysticism might be characterized as the study of those propositions which are equivalent to their own negations. The Western point of view is that the class of all such propositions is empty. The Eastern point of view is that this class is empty if and only if it isn't.*
>
> RAYMOND SMULLYAN[7]

The conventional wisdom is that systems of reasoning must be consistent. That is, no statement can be both true and false. If so, then the system collapses because there remain no restrictions on what is true or false: every statement can be proved true (and false as well!). When Bertrand Russell once made this claim during a public lecture he was challenged by a sceptical heckler to prove that the questioner was the Pope if twice 2 were 5. Russell replied, 'if twice 2 is 5, then 4 is 5, subtract 3; then $1 = 2$. But you and the Pope are 2; therefore you and the Pope are one.'!

An interesting feature of this situation is that it shows us how far our minds are from being like a computer when they reason (as opposed to how they operate at a neurological level). Each of us holds all sorts of contradictory views which would render us inconsistent if we were computing machines; none the less, we do not believe that every statement is true.

We have seen that Tarski's analysis of the hierarchy of statements and metastatements removes one collection of apparent contradictions which threatened to bring the whole edifice of logic crashing down. It is interesting to inquire what might be the physical analogue of a contradiction, and to ask whether Nature might be inconsistent in some way, or be describable (in whole or in part) by a mathematical system that is inconsistent.

The consistency of Nature must mean that in some sense there are no true paradoxes (or, more weakly, perhaps, that none are observable). Human attitudes to this assumption have an ancient and unusual history. One finds many examples of contradictory ideas living together, accommodated by the addition of some principle of complementarity. Indeed, many religions make great play of such aspects ('I am Alpha and Omega') as a way of affirming and reinforcing the transcendental nature of the Deity.[8] This places many aspects of the Deity's existence and nature beyond the reach of human reasoning and scepticism. However, it is also the case that the great monotheistic faiths provided some basis for the rationality of Nature. They viewed the world as an outworking of the mind of a rational Creator and so its consistency was expected. Yet, the possibility of chaos and irrationality, either in the past, before the creative intervention of the Deity, or at some time in the future, was often another part of the story. Another complication in this picture is that miracles are usually countenanced in these religious systems. But they need not produce events which would be judged inconsistent with the normal course of events— although they might.

Some of the earliest attempts to grapple with logical problems like this, which go to the heart of our contemplation of the rationality of the Universe, arise in medieval Christian theology. These attempts confront the thorny problem of whether God can alter the past. If he can change the past, then the moral and rational order of things is turned upside down; if not, does this then compromise God's omnipotence? The most prominent supporter of the view that the past could be changed was an eleventh-century Italian, St. Damian. He maintained that God's power was not bounded by time and that 'God can make it so that Rome, even after it was founded, should not have been founded.'[9]

Ranged against this radical view, we find the contrary opinion espoused most thoroughly by Thomas Aquinas two hundred years later. Aquinas's conception of God required that there be no contradictions, and that God be bound by his own laws. As we discussed in Chapter 1, this led to the notion that God can do everything that can be done (rather than everything that we can conceive of). Other commentators were more explicit, constraining God's actions even further within the set of all things that could be done, by restricting his actions to right actions. This is the position that Milton adopts in *Paradise Lost*. Right actions not only included those which were morally good, but they possessed logical consistency and rationality as necessary properties.

These debates were the start of debates over the theology of changing the past, or praying for such changes to occur, which continue to this day. Is it possible to pray for the past to be changed? Few Christian theologians would support this idea if the past was known to the person praying; but what if an event has occurred about which the outcome is still unknown to you? Or what about an outcome, like an examination result, that has already been decided, but which has yet to be announced? In his book on miracles,[10] C.S. Lewis, an influential popular writer on theological questions, sided with the idea that it was rational to pray for events whose outcome had already been decided, because from a God's-eye perspective your future intercession could be an ingredient in the global events which may affect the outcome of the event being prayed for. Lewis was adopting what physicists call the 'block universe' picture of spacetime, in which the entire spacetime already exists as a complete entity.[11] He conceived of the whole of space and time as viewed externally by God, and so all prayers were known by God before they were made. This would permit free will to be retained together with a doctrine of God's omniscience. God's foreknowledge does not predestine our actions. Rather, it is our actions that determine God's foreknowledge. We introduce these interesting theological questions to show that the questions of changing the past and making sense of the resulting coherence of the Universe are not questions that lie solely in the realm of physics.

We have introduced the idea that Nature might display inconsistency. What might we mean by an inconsistency of Nature? How would we recognize one when we saw it? At first, it seems that such things are inconceivable unless there is some gross mistake in our formulation of Nature's laws. But things might not be quite so simple.

Time travel: is the Universe safe for historians?

The secret of time travel may be discovered by physicists, but its use as a weapon will be decided by historians.

PAUL NAHIN[12]

Einstein's theory of gravitation—the so-called general theory of relativity—is the most accurate scientific theory that we possess. It predicts the changes observed in a distant pulsar with an accuracy of one part to 10^{14}. No observation has ever been made anywhere in Nature which conflicts with the predictions of this theory. However, along with all the successful accounts of the things we see in the Universe, like black holes, neutron stars, gravitational lenses, and the idiosyncrasies in the motions of the planet Mercury, come other more puzzling possibilities. The American physicist Michio Kaku writes that

> Einstein's equations, in some sense, were like a Trojan horse. On the surface, the horse looks like a perfectly acceptably gift, giving us the observed bending of starlight under gravity and a compelling explanation for the origin of the universe. However, inside lurk all sorts of strange demons and goblins, which allow for the possibility of interstellar travel through wormholes and time travel. The price we had to pay for peering into the darkest secrets of the universe is the potential downfall of our most commonly held beliefs about our world—that its space is simply connected and its history is unalterable.[13]

In 1949, the logician Kurt Gödel,[14] of whom we shall hear much more in what follows, discovered that Einstein's theory allows time travel to occur. It is still not known whether this could be realized in our particular universe.[15] In fact, writing as early as 1921 in his famous text *Space, Time, and Matter*,[16] the great mathematician and physicist Hermann Weyl wrote about this possibility with extraordinary prescience, nearly thirty years before Gödel:

It is possible to experience events now that will in part be an effect of my future resolves and actions. Moreover, it is not impossible for a world-line (in particular, that of my body), although it has a time-like direction at every point, to return to the neighbourhood of a point which it has already once passed through. The result would be a spectral image of the world more fearful than anything the weird fantasy of E.T.A. Hoffmann [a nineteenth-century author of fantastic literature] has ever conjured up. In actual fact the very considerable fluctuations of the [metric tensor of spacetime] that would be necessary to produce this effect do not occur in the region of the world in which we live. Although paradoxes of this kind appear, nowhere do we find any real contradiction to the facts directly presented to us in experience.

It is interesting to note that Weyl was a member of the Institute of Advanced Study in Princeton during the period when Einstein and Gödel were there. They would no doubt often have spoken together about unsolved scientific problems (Einstein and Gödel talked together almost daily). Perhaps Weyl influenced Gödel's thinking about time travel in some way?

If time travel can occur, then we seem to be facing inconsistency in Nature. It looks as if we could create factual contradictions by changing the past in ways that could not give rise to the present. You could bring about the death of your ancestors so as to exclude the possibility of your own birth. Your current existence would then seem to constitute a logical contradiction. We could also create information out of nothing. I could learn Pythagoras' theorem from a mathematics textbook today, then travel back in time to meet Pythagoras as a young man in order to give him the idea for his theorem before he had thought of it. Where would the information in the theorem have come from? I learned it from the inheritance of Pythagoras; but Pythagoras learned it from me! Suddenly we are in the world of the *X-files*. The possibilities are endless. Time travel is the thinking man's UFO. Ponder this little story by Dennis Piper; it is called *The Oscillating Universe*:

> One day the Professor called me in to his laboratory. 'At last I have solved the equation,' he said. 'Time is a field. I have made this machine which reverses this field. Look! I press this switch and time will run backwards run will time and switch this press I. Look! Field this reverses which machine this made have I. Field a is time,' said he, 'Equation the solved have I last at.' Laboratory his to in me called Professor the day one. 'For heaven's sake, SWITCH IT BACK,' I shouted. *Click!* Shouted I, 'BACK IT SWITCH, sake heaven's for.' One day the Professor called me into his Laboratory . . .

Logical paradoxes of the 'what-if-I-killed-my-grandfather' type constitute a *genre* called 'Grandfather Paradoxes' by philosophers interested in time travel. They appear to beset any form of backward-in-time travel (as opposed to forward-in-time travel), and have been a prominent component of science fiction stories about time travel ever since the scenario of machine-borne time travel began with H.G. Wells's 1895 classic, *The Time Machine*, which appeared in serial form in the *New Review*. More recently, it formed the central plot of the 1984 film *The Terminator* (and sequels 2, 3,... $N \rightarrow \infty$) in which a time-travelling robot from the year 2029 turns up in present-day Los Angeles (where else?) in order to murder the woman who will give birth to a son who has made enemies in the future. Just one year later, another film, *Back to the Future*, explored a similar problem with more comic intentions. The hero, Marty McFly, travels back to the 1955 high-school days of his parents and, by accident, almost prevents them from marrying. He even has the shocking experience of seeing his own image fading out of the family photograph that he carries around in his wallet. In Britain, *Doctor Who* is the long-running cult TV series. The 'Doctor' travels through time with young companions in a converted police telephone box of a classic external design, alas no longer visible in Britain.

There is a wide diversity of opinion about the conclusions to be drawn from these mind-bending possibilities. Some regard the Grandfather Paradoxes as a proof that time travel is forbidden in our Universe (a weaker version of this prohibition would allow time travel in so far as it did not create changes in the past). For example, the well-known science fiction writer Larry Niven wrote an essay in 1971 entitled 'The theory and practice of time travel' in which he enunciated 'Niven's Law' of time travel: '*If the Universe of Discourse permits the possibility of time travel and of changing the past, then no time machine will be invented in that Universe*'. Niven is convinced that time travel is equivalent to the introduction of irreconcilable inconsistency in the Universe and must be prohibited by some self-consistency principle deep within Nature's laws. Since no such exclusion principle has been found, we must introduce it by fiat.

Nor are such worries confined to science fiction writers. In 1992, the physicist Stephen Hawking gave the same general 'no time travel' idea the name of the *Chronology Protection Conjecture*.[18] Hawking believes that time travel into the past cannot be possible because 'we have not been invaded by hordes of tourists from the future' arriving to watch or change great moments in history. But we might well ask how we would know what to look for, or how we would tell whether the 'normal' course of history was being disrupted by time travellers: perhaps J.F. Kennedy would have started World War III in 1964 if he had not been assassinated a year earlier?

The Chronology Protection Conjecture asserts that the laws of physics prevent the creation of a time machine, except at the beginning of time (when there is no past to travel into). Its purpose is to stimulate physicists to investigate Einstein's equations, and those equations which define the new superstring 'theories of everything', to discover the conditions under which the conjecture is true.

This 'tourists from the future' problem has a long history. It is known as the 'cumulative audience paradox' among science fiction writers, after Robert Silverberg's explicit introduction of it in 1969. As time travellers flock to the past, the worry is that an ever-increasing number of people accumulate at significant events in our history. Silverberg argues that events like the Crucifixion would attract billions of time travellers, yet 'no such hordes were present' at the original event. More generally, we shall find our present and past increasingly clogged with voyeurs from the future: 'A time is coming [when time travellers] will throng the past to the choking point. We will fill our yesterdays with ourselves and crowd out our own ancestors'[19]. These visitors would, in effect, be gods: they would have control over time and access to all knowledge. Perhaps the level of technical knowledge that makes such travel possible also reveals the deep problems that its exploitation would create, and wisdom ensures that the knowledge is never exploited. Time travel offers the possibility of destroying the coherence of the Universe in the same way that our knowledge of nuclear physics offers us the means of destroying the Earth. In fact, John Varley in his science fiction story *Millennium* (1983), is worried that

> Time travel is so dangerous it makes H-bombs seem perfectly safe gifts for children and imbeciles. I mean, what's the worst that can happen with a nuclear weapon? A few million people die: trivial. With time travel we can destroy the whole Universe, or so the theory goes.[20]

Fortunately, having that knowledge does not require its practical use. To a considerable extent, maturity, whether it is in young people, nations, or entire civilizations, is closely associated with the growth of self-restraint: that is, with a growing recognition that not everything that you can do, or want to do, should be done.

This type of 'where are they?' argument against time travellers is rather reminiscent of Enrico Fermi's famous 'where are they?' response to claims for the existence of advanced extraterrestrials.[21] Some possible reasons for the absence of advanced extraterrestrials are:

(1) There aren't any yet able to signal. We are the most advanced life form within communicating range.

(2) Technological civilizations cannot survive for long enough to become super-advanced. They blow themselves to bits, are wiped out by asteroidal impacts, or succumb to other internal problems—disease, exhaustion of raw materials, or irreversible degeneration of their environment by pollution.

(3) There are so many civilizations, and ours is a fairly average example of which there are millions of others. Therefore the most advanced extra-

terrestrials have no reason to take any special interest in us. We are just like another species of common insect.

(4) Advanced extraterrestrials have a rigid code of non-interference in the histories of more primitive civilizations. We are like a cosmic game reserve; we are being studied but in a non-intrusive fashion.

(5) Advanced extraterrestrials exist, but communicate only with technology at levels exceeding our own. In this way they require that a particular level of scientific maturity is required before any civilization can join the 'club'.

Each of these responses can be applied to the problem of why there are no time travellers. But in the case of time travel there is the possibility of a fundamental self-prohibition being imposed by super-advanced extraterrestrials because they understand more fully that there would be grave consequences for the coherence of the whole of space–time if time travel were indulged in. There is no real analogue of this in the case of communication with advanced extra-terrestrials.

The most novel version of the 'where are they?' argument must surely be that proposed by the economist M.R. Reinganum, who wrote an article with the title 'Is time travel possible?: a financial proof'.[22] He argues that the fact that we see positive interest rates proves that time travellers do not exist. (He also claims that it means they cannot exist, which does not follow at all.) The reasoning is simply that time travellers from the future could use their knowledge to make such huge profits all over the investments and futures markets that interest rates would be driven to zero. This argument reminds me of one against claims of clairvoyance: psychic powers would give their possessors the ability to get rich from any form of gambling. Why bother bending spoons and guessing cards when you can win the National Lottery every week? If these powers existed in humans then they would have bestowed such advantages upon their recipients that they would have become dominant in many ways, and the ability should have evolved throughout successful human sub-populations.

Another set of responses to the time-travel paradoxes is to assert another principle, less sweeping than the exclusion of all time travel proposed by Niven or Hawking, which simply requires the immutability of the past in order to guarantee our immunity from paradox and contradiction. Perhaps our scope for action is strongly limited by consistency, and you can't alter the past without altering the present too. Or perhaps a principle of the conservation of reality is so powerful that no tinkerings with the past are possible, especially the act of travelling there.

There might, of course, be rather more mundane restrictions which simply render time travel uneconomic. Tourism from the future might require such enormous energy expenditure that the whole idea is always hopelessly impractical, even if it is possible in principle. Gödel himself offered only this mundane practical defence against the paradoxes of time travel in his original article. Having recognized that in his newly discovered universe model

> it is possible in these worlds to travel into any region of the past, present, and future, and back again, exactly as it is possible in other worlds to travel to distant parts of space. This state of affairs seems to imply an absurdity. For it enables one e.g., to travel into the near past of those places where he has himself lived. There he would find a person who would be himself at some earlier period of his life. Now he could do something to this person which, by his memory, he knows has not happened to him . . . This and similar contradictions, however, in order to prove the impossibility of the worlds under consideration, presuppose the actual feasibility of the journey into one's own past. But the velocities which would be necessary in order to complete the voyage in a reasonable time are far beyond everything that can be expected ever to become a practical possibility [they must exceed 71 per cent of the speed of light]. That is, it cannot be excluded a priori, on the ground of the argument given, that the space–time structure of the real world is of the type described.[23]

But not everyone is persuaded by these arguments against the possibility of time travel. There is something not quite coherent about all these arguments about *changing* the past. The past was what it was. You cannot alter it and expect the experienced present still to exist. We might have been there influencing it; but how could there be two pasts: one which was, and another which would have been if we had intervened, but which are in some way inconsistent with one another? If you could travel back in time to prevent your birth, then you would not be here to travel backwards in time for that purpose.

The American philosopher David Malament discusses this common view that, because of Grandfather Paradoxes,

> time-travel . . . is simply absurd and leads to logical contradictions. You know how the argument goes. If time travel were possible, one could go backward in time and undo the past. One could bring it about that both conditions P and not-P obtain at some point in spacetime. For example, I could go back and kill my earlier infant self, making it impossible for that earlier self ever to grow up to be me. I simply want to remark that arguments of this type have never seemed convincing to me . . . The problem with these arguments is that they simply do not establish what they are supposed to. To be sure, if I could go back and kill my infant self, some sort of contradiction would arise. But the only conclusion to draw from this is that if I tried to go back and kill my infant self then, for some reason, I would fail. Perhaps I would trip at the last minute. The usual arguments do not establish that time travel is impossible, but only that if it were possible, certain actions could not be performed.[24]

This argument is interesting but not entirely convincing. The possibility of time travel seems to be allowed by the laws of gravitation physics. It would seem very strange if the protection from self-contradiction has to appeal to something completely outside the realm of the laws of gravitation—like accidents of history—which stop otherwise routine situations arising because they would lead to paradoxes in the future. Perhaps the allowed time-travel trips have such enormous lengths that they are all longer than the age of the Universe and so can have no practical effect upon it.

What is missing from the arguments against time travel is well illustrated by a situation that arose in the 1970s in the study of black holes. Roger Penrose had proposed that places where the laws of physics break down—singularities in spacetime—could occur only inside black holes where they are shielded from the outside world by the boundary surface of the black hole's horizon. Since nothing can pass out through the horizon to the outside universe, we are protected from the unpredictable outpourings of the singularity at the heart of the black hole. This idea was called the Cosmic Censorship Hypothesis, and we discussed it in the previous chapter. It is not dissimilar to the Chronology Protection Conjecture in its protectionist aim of circumscribing irrationality. Of course, a crucial aspect of this problem is whether such a singularity really forms anywhere (even at the centre of a black hole) when all the laws of physics (especially quantum gravity) are included. But, ignoring quantum theory, it is a challenging question to decide whether a naked singularity (that is, one that is not shrouded by a horizon) can form in the Universe. In contrast to the search for protection from time-travel paradoxes, it was found that gravity conspired to stop singularities from forming in subtle ways. For example, it was known that if a black hole could be made to rotate faster than some critical rate, then the horizon would shrink to zero size and outside observers like us could see and be affected by the central singularity. So, suppose that we begin with a black hole that is rotating very fast, but just a little slower than the critical rate needed if we are to see the singularity. Now, drop a body, rotating in the same directional sense, into the black hole so that the slightly enlarged black hole will have more rotation than the critical value.

However, as this simple thought experiment was explored in more detail, a remarkable situation came to light. In Einstein's theory of gravitation, there exist forms of gravitational force that are not present in Newton's theory. One of these forces is a repulsive force between bodies rotating in the same sense. It turned out that whenever the infalling spinning body was rotating in such a way that its capture would result in a black hole spinning faster than the critical level, the repulsive spin force was sufficient to stop the body entering the black hole. This was a remarkable example of Nature's cosmic censorship at work. In the case of time travel, there is no trace of any such physical mechanism emerging to protect chronology and historical determinism arising within the same physical theories which allow time travel.

Another distinguished philosopher who has swum against the tide and has argued for the rationality of time travel in the face of the Grandfather paradoxes is David Lewis. In 1976, he wrote that

> Time travel, I maintain, is possible. The paradoxes of time travel are oddities, not impossibilities. They prove only this much, which few would have doubted: that a possible world where time travel took place would be a most strange world, different in fundamental ways from the world we think is ours.[25]

If we look more closely at the logic of the Grandfather Paradoxes we see that there is a nagging problem about their coherence that Malament and Lewis worry about. Time travel must not involve undoing or changing the past in a manner that implies that there are two pasts: one without your intervention and one with it. If you travel back to influence some historical event, then you would have been part of that event when it occurred. A contemporary historical record would have included your presence (if you were noticeable). Time travellers do not change the past because they cannot do anything in the year 1066 that was not actually done in 1066. Someone can be present at an event in the past and contribute to the record of what happened in history; but that is quite different from the presumption that they can change the past. If a change occurs we can ask for the date when that change occurred. In the same way, the philosopher Larry Dwyer has argued that

> Time travel, entailing as it does backward causation, does not involve changing the past. The time traveller does not undo what has been done or do what had not been done, since his visit to an earlier time does not change the truth value of any propositions concerning the events of that period . . . It seems to me that there is a clear distinction to be made here, between the case where a person is presumed to change the past, which indeed involves a contradiction, and the latter case where a person is presumed to affect the past by dint of his very presence in that period.[26]

A further worry about the analysis of time travel is the exclusion of any considerations of quantum mechanics. Our experience of the quantum theory is that it is far less restrictive about what can happen in the world than is non-quantum physics. Instead of telling us that a given cause has a particular effect, it tells us only that there is a whole array of different possible outcomes with different probabilities. In some cases, as when you drop a glass on the floor, the probability is overwhelmingly dominated by one type of outcome: the one that the Newtonian physicist would expect to see. However, in the subatomic realm these probabilities can be more evenly spread among different possibilities, and we can see different outcomes from the application of identical causes.

Quantum mechanics excludes very little. Almost anything can happen with some probability, but the probabilities of witnessing things that we would be tempted to call miracles (like human levitation) are so low that they would not be likely to be witnessed in our vicinity of space during the whole history of the human race. David Deutsch has grappled with these quantum time-travel problems, and has argued that we should take the possibility of time-travelling paths very seriously if we want to arrive at the correct formulation of the quantum theory of gravity.[27] When we compute the probabilities that certain changes will occur to the Universe we can either include or ignore the time-travelling paths when calculating what the actual probabilities are. What we would like to find is an experimental situation in which the relative frequencies of different outcomes were different when time-travelling paths were included. This would provide an experimental test of whether these paths are present in reality or not. Deutsch gives a possible quantum resolution of the Grandfather Paradoxes, which shows that the 'quantum universe' to which the time traveller returns is never the same (in Everett's 'many worlds' sense). Deutsch does not like the idea of getting information for nothing by travelling into the past with your current knowledge in order to 'invent' something. Clearly, the entire theory of the evolution of life by natural selection could be circumvented by this means: organisms could be trained or forewarned of hazards that they must overcome later in their evolutionary history. Following the tendency to invoke 'principles' to legitimize these uncomfortable feelings about time travel, he proposes a 'principle of evolution' to prohibit getting information for nothing by time travel.[28]

Completeness

I don't believe in mathematics.

ALBERT EINSTEIN

At the end of the nineteenth century, the *fin de siècle* urge for completion was prominent in the world of mathematics. It was not a case of mathematicians thinking that they would be able to discover all the mathematics that there was. They knew that could never be done. Rather, there was a movement headed by David Hilbert, the greatest mathematician of the day, which sought to get a grip on the problems of paradox that had been surfacing. How could you be sure that mathematics was a reliable system of reasoning? If someone gave you a statement about numbers, how would you set about demonstrating whether it was true or false? Hilbert challenged mathematicians to come up with a recipe by which the truth of any statement of mathematics could be tested. If all statements can be tested in this way, the logical system to which they belong is called *decidable*. If all truths in its language can be deduced from its axioms, it is called *complete*. As an illustration, consider a board game like chess or Go.

Incompleteness would mean that there were configurations of pieces on the board that could not have been reached from the starting layout by following the rules of the game.

When Hilbert began to worry about the challenges to the edifice of mathematics posed by simple logical paradoxes, he was seeking a guarantee that arithmetic was consistent: that is, it could not be shown that $0 = 1$. Yet, Hilbert saw consistency as more than an insurance policy on the health of the body of mathematics. He wanted to use it as the definition of mathematical truth. Any consistent statement 'exists' as a mathematical statement. This makes mathematics something rather larger than the physical world of experience, for not everything that can exist mathematically appears to exist in physical reality.

Hilbert began by carrying through this programme for a famous logical system, well known to generations of school children: Euclid's geometry, with its points, lines, and angles. When all the axioms and rules of reasoning were laid out in full, Hilbert showed that Euclidean geometry was a complete, decidable logical system: any statement about points, lines, and angles could be shown to follow from the axioms after a number of deductive steps or, if not, be shown to lead to a contradiction, and thus to be false.

Hilbert and others carried this programme a little further by proving that some other simple logical systems were complete. Hilbert was using these studies as a warm-up exercise for an attack upon the big prize: a proof that all statements of *arithmetic* can be judged true or false. This is a much harder problem because arithmetic is a much larger and richer system than Euclidean geometry. But Hilbert aimed to attack it by enlarging the complexity of each system whose completeness he proved, expecting that, step by step, he would eventually take the fortress of arithmetic.

Hilbert's quest for completeness has interesting links to the general end-of-the-century pessimism about the scope of science, which we have seen was voiced by Emil Du Bois Reymond and his allies. Hilbert wanted to assert the limitless power of mathematics to decide whether statements of mathematics were true or false. He was convinced that

> every definite mathematical problem must necessarily be susceptible of an exact settlement, either in the form of an actual answer to the question asked, or by the proof of the impossibility of its solution and therefore the necessary failure of all attempts to solve it.

Whereas Du Bois Reymond's rallying cry had been *Ignoramus et ignorabimus* ('We are ignorant and we shall remain ignorant'), Hilbert issued his challenge to mathematicians to ignore this call, claiming that

> we hear within us the perpetual call. There is the problem, Seek its solution. You can find it by pure reason, for in mathematics there is no ignorabimus.

His other *bête noire* was Comte, whose views we have also seen in Chapter 2; he claimed that

> The true reason, according to my thinking, why Comte could not find an unsolvable problem lies in the fact that there is no such thing as an unsolvable problem.

In his most famous lecture, to the International Congress of Mathematicians in 1900, where he listed what in his opinion were the greatest unsolved problems for mathematicians of the twentieth century to solve, Hilbert returned again to this deep issue of solvability and its links to the nature of mathematics and to the human mind, asking,

> Is this axiom of the solvability of every problem a peculiarity characteristic only of mathematical thought, or is it possibly a general law inherent in the nature of the mind, a belief that all questions which it asks must be answerable by it? For in other sciences also one meets old problems which have been settled in a manner most satisfactory and most useful to science by the proof of their impossibility. I cite the problem of perpetual motion. After seeking unsuccessfully for the construction of a perpetual motion machine, scientists investigated the relations which must subsist between the forces of nature if such a machine is to be impossible; and this inverted question led to the discovery of the law of the conservation of energy, which, again, explained the impossibility of perpetual motion in the sense originally intended.

Although it was clear that arithmetic was a bigger system that geometry, it was not obvious that there was anything intrinsically different about it. Yet Hilbert was unable to extend his elegant proofs of the completeness of geometry to larger systems; later, he would discover why it was proving so difficult. In 1930, Kurt Gödel (the very same Gödel who, 19 years later, would surprise the world of science with the discovery that Einstein's equations allowed time travel: see p. 199) announced a sensational and completely unexpected result: that any logical system rich enough to contain arithmetic must be either incomplete or inconsistent. There must exist statements of arithmetic whose truth or falsity cannot be established using the axioms and deductive rules of arithmetic. Arithmetic truth is something too large to be ensnared by any formal system of rules. This result was to change the way we think about human reasoning and it would be the choice of many for the accolade of the most important mathematical discovery ever made. For the first time, logic proved that there were things that could not be proved. Logic had become a true religion.

One should not misunderstand what Gödel's theorem says about the consistency of arithmetic. It does not say that we cannot prove the consistency of arithmetic. We can. But our proof cannot be formalized within the language of arithmetic. Gödel's theorem simply says that arithmetic cannot prove the consistency of arithmetic. Indeed, if arithmetic had been able to prove its own consistency, the situation would not have been entirely convincing. It would be like the police investigating complaints against the police, or a politician arguing that he should be believed because he is truthful.

For more than forty years, the impact of Gödel's discovery was entirely negative. It showed that Hilbert's goal could not be realized: mathematics was mysteriously open-ended. Gödel went on to show that the self-consistency of the axioms of systems as complex as arithmetic could not be demonstrated either. Subsequently, some of the wider implications of Gödel's result started to become evident as his proof was recast into other forms which led to further powerful results.

Alan Turing was the first person to contemplate the capabilities of automatic calculating machines—what we would now call computers. Defining such machines simply by their ability to read a list of numbers and change them into another list, he wanted to know what their ultimate capabilities were. Could they compute any mathematical formula you cared to invent? No! Arguments like Gödel's displayed mathematical questions which the 'computer' could never answer in any finite series of operations: there were mathematical problems which would take the computer for ever to solve. This general problem of determining whether or not a computer calculation would ever halt has become known as the 'halting problem'. Those mathematical questions which can be decided in a finite number of computational steps are called 'computable', the rest are called 'uncomputable'.

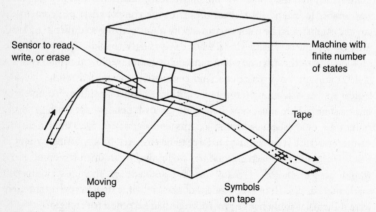

Fig. 7.3 Turing's idealized machine, first conceived in 1936. It consists of a finite collection of symbols, a finite number of different states in which the machine can reside, an endless tape marked with slots, each of which carries a single symbol, and a sensor which scans, reads, and writes on the tape, together with a set of instructions giving the rules that cause changes (or no changes) to be made to the tape after a slot is read.

In order to capture the capabilities of any mechanical calculating device, Turing invented an idealized computing device. (This was before the days of computers as we know them: the word 'computer' then just meant a human calculator.) His device became known as a 'Turing machine' (Fig. 7.3).

Uncomputability usually has a negative impact upon those who learn of it; but that reaction is far from universal amongst mathematicians. They see the open-endedness of mathematics as being a wonderful thing. G.H. Hardy wrote of his distaste for a mathematical world in which Hilbert's goal could be achieved:

> Suppose, for example, that we could find a finite system of rules which enabled us to say whether any given formula was demonstrable or not. This system would embody a theorem of metamathematics. There is of course no such theorem, and this is very fortunate, since if there were we should have a mechanical set of rules for the solution of all mathematical problems, and our activities as mathematicians would come to an end.

Impossible constructions

Most architects think by the inch, talk by the yard, and should be kicked by the foot.
PRINCE CHARLES

There are many areas of human activity where we set ourselves exercises which must be achieved in the face of some constraint. Blindfold chess, race-walking, and handicapped horse racing are all examples of activities where some deliberate constraint is applied in order to make a goal more challenging to achieve. Sports are not alone in liking this formula for making life more interesting. Mathematicians have long had an interest in finding out whether it is possible to do things using only particular tools. The Greeks' love of practical geometry led them to investigate what things were possible by the use of a straight edge (sometimes referred to as a 'ruler', but no use can be made of the markings to measure lengths) and compasses alone. A ruler enables you to draw a straight line between two points; a pair of compasses enables you to draw an arc or a circle and to mark out equal distances. These were the basic tools of architects at the time, and this whole problem clearly had the serious practical purpose of discovering the procedures they should follow to carry certain routine constructions when drawing up their building plans. Indeed, one can find essentially the same problems posed and solved (in the same way) in other advanced ancient cultures, like that of early India, where these constructions were required for the construction of altars and for religious ceremonies.[29]

Consider the simplest problem of this sort: how do you bisect a line? The midpoint of any line can be found using the compasses by drawing two arcs, one centred on each end of the line (of radius greater than half the length of the line, so that they intersect). Now draw a straight line between the two points where the two arcs have intersected each other. It passes through the midpoint of the line (Fig. 7.4).

The next step after the bisection of a line is to ask if it is possible to bisect an angle with these tools (Fig. 7.5). Draw any angle, then put the compass point at the corner of the angle, A; next, draw any arc that cuts the two lines. Now we just need to find the midpoint of the arc joining these two intersections. Draw two arcs, centred at B and C; then draw a line with the rule from the point where these arcs intersect to the corner A. This line bisects the angle.

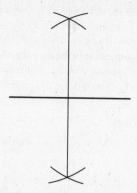

Fig. 7.4 Dividing a line in half by ruler and compass' constructions. Set the pair of compasses to an arc of radius greater than half the length of the horizontal line. Draw a circle of this size centred on each of the end points of the line. They will intersect at two points, above and below the line. The vertical straight line joining these two points divides the horizontal line in half.

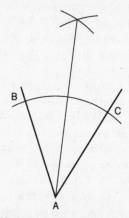

Fig. 7.5 Dividing an angle in half. Place the compass point at the corner of the angle, A. Draw any arc that intersects both lines forming the angle at B and C. Now draw two arcs of equal radius from each of the two points of intersection. The straight line from the point where these two arcs intersect to the corner of the angle divides the angle in half.

All this was child's play to the early Greek geometers. They kept exploring the scope of their method of 'ruler-and-compass construction', firm in the belief that anything could be achieved if one was ingenious enough. Their interest crystallized around one problem that they could not crack: how to *trisect* an angle. This problem remained unsolved until 1837, when Pierre Wantzel proved that such a ruler-and-compass construction is impossible. Curiously, Wantzel remains virtually unknown as a mathematician, and received surprisingly little acclaim for solving a two-thousand-year-old problem even in his own day. Wantzel achieved his proof by changing the problem into one of algebra. Important developments had been made in this subject by Ruffini and Abel, which Wantzel used to establish the impossibility of trisection. Mathematicians viewed the algebraic developments as being deeper and more fundamental than their application to the trisection problem and, as a result, Abel became far more renowned for his role in establishing this field. Ruffini and Abel showed that no algebraic equation of degree greater than four allows us to find its solutions by a formula. Again, this problem had some history.[30] The solution for equations of degree one was trivial; degree two had been known for thousands of years; degrees three and four had been solved by the Renaissance mathematicians Scipio del Ferro, in 1515, and Ferrari, in 1545. Competition had been intense amongst mathematicians ever since to solve the next case—for, perhaps some clever rule might solve them all at one fell swoop.

Abel, aided by the work of Galois, finally established an impossibility theorem. Later, he discussed the general question of solubility in mathematics and, sounding a little like Hilbert many years later, he realized that any attempt at complete understanding of mathematical problem must have two means of attack; one to find explicit solutions; the other, to discover whether solution is possible or not. Only in this way could a problem be closed, because

> To arrive infallibly at something in this matter, we must therefore follow another road. We can give the problem such a form that it shall always be possible to solve it, as we can always do with any problem. Instead of asking for a relation of which it is not known whether it exists or not, we must ask whether such a relation is indeed possible . . . When a problem is posed in this way, the very statement contains the germ of the solution and indicates what road must be taken; and I believe there will be few instances where we shall fail to arrive at propositions of more or less importance, even when the complication of the calculations precludes a complete answer to the problem.[31]

Interestingly, Abel at first thought he had found the solution for the degree five problem. But before his paper could be published he found a mistake and, as a result, started to see the problem in a completely different light. This vital change of perspective led ultimately to his proof of the impossibility of the very result that he thought he had once established.

It appears that Abel's work did not give rise to deep philosophical and theological speculations about why it was that solubility stopped at degree 4. Clearly, it could have done. Equations of higher degree certainly have solutions. We can solve *some* of them by inspired guesswork, approximations, and so forth (as could mathematicians in Abel's day) but Abel's proof seemed to open up a gap between what human reasoning could achieve and what was true in the transcendental world of mathematical truths, or in the mind of God.

Many of the philosophical issues raised by Gödel's theorem could have been stimulated by these discoveries that there are limits to our ability to solve algebraic equations and to the scope of ruler-and-compass construction, but they were not. There are many analogies between the two lines of inquiry. Both Abel and Gödel attacked problems that everyone expected could be solved. Both displayed remarkable flexibility of mind in establishing an impossibility theorem: Abel did a last-minute about-turn after thinking he had got a 'possibility' theorem, and Gödel had actually been proving the completeness of smaller logical systems than arithmetic (this was his doctoral thesis work) just months before announcing his impossibility theorem for arithmetic.

Gödel established a correspondence between statements of mathematics and statements about mathematics (metamathematics). He did this by using prime numbers to encode each ingredient of a logical or mathematical statement. The product obtained by multiplying the prime numbers together then defines the whole statement. This number is now called its Gödel number. Moreover, since any number can be expressed as a product of prime numbers in one and only one way (for example, $51 = 3 \times 17$, $54 = 2 \times 3^3$, $9000 = 2^3 \times 3^2 \times 5^3$) the correspondence is unique: to each Gödel number there corresponds a logical statement. In this way every Gödel number corresponds to some logical statement about numbers (not necessarily a very interesting one) and each statement about numbers corresponds to some Gödel number.[30] For example, the Gödel number $243,000,000 = 2^6 \times 3^5 \times 5^6$. The logical sentence is defined by the powers of the prime numbers taken in order, that is 656. The symbol 6 corresponds to the arithmetic object zero, 0, while 5 corresponds to =, and so this Gödel number represents the rather uninteresting arithmetical formula $0 = 0$.

Gödel decisive step is to consider the statement

The theorem possessing Gödel number X is undecidable.

He calculates its Gödel number and substitutes that value for X in the statement. The result is a theorem that establishes its own unprovability.[33]

The essential feature that make the incompleteness argument work is the possibility of self-reference: the correspondence between arithmetic and statements about arithmetic. This is possible only in logical systems which are complicated enough to allow statements about them to be coded uniquely and completely within the systems themselves, so that if each possible ingredient of a logical statement is ascribed to a different prime number, then any complete statement can be represented by a Gödel number which can be factorized uniquely to give the statement about arithmetic to which it corresponds. Some logical theories, like geometries, do not contain enough machinery to allow statements about themselves to be encoded within them in this way. These theories cannot display incompleteness.

Metaphorical impossibilities

No non-poetic account of reality can be complete.

JOHN MYHILL

Some attempts have been made to create a metaphorical extension of the insights of Gödel and Turing in order to reveal what sorts of things in our experience might be outside the grasp of formalisms and limits. In 1952, the American logician John Myhill produced the most interesting of these.[34] Myhill classified possibilities by borrowing some terminology from mathematical logic and dividing ideas into three classes: 'effective', 'constructive', and 'prospective'.[35]

The most accessible and quantifiable aspects of the world have the attribute of computability. There exists a definite mechanical procedure for deciding if anything does or does not possess a particular property. Computers and human beings can be trained to respond to the presence or absence of such attributes. Truth is not a computable property, while being a prime number is.

A more elusive set of attributes are those which are merely listable. For these, we can construct a procedure which will list all the cases that possess the desired attribute (although you might need to wait for an infinite time for the listing to be completed). There is, however, no way of producing a listing of all the cases that do not possess the attribute. Many logical systems are listable, but not computable: all their theorems can be listed, but there is no mechanical procedure for deciding whether any given statement is or is not a theorem. In a mathematical world with no Gödel theorem, every statement would be listable. In a world without Turing's uncomputable operations, every property of the world would be computable.

The problem of deciding whether this page possesses the attribute of grammatically correct English is a computable one. But the page could still be meaningless to a non-English reader. With time, the reader could learn more and more English, so that bits of the page became intelligible, but there is no way of predicting where the meaningful parts will be located on the page. The attribute of meaningfulness is thus listable but not computable.

Not every attribute of things is listable or computable. The property of being a true statement of arithmetic is an example. Attributes which are neither listable not computable are called 'prospective': they can be neither recognized nor generated in a finite number of deductive steps. They show that there is a place for ingenuity and novelty. There are things which cannot be encapsulated by any finite collection of rules or procedures. Beauty, simplicity, ugliness, and truth are all prospective properties. There can be no magic formula that can generate all possible examples of attributes like these, even in an infinite life-time. They are inexhaustible. No program or formula can generate all examples of beauty or ugliness; nor can any program recognize them all when it sees them, and nor can we, in the way that the romantics imagined. In Myhill's words, 'The analogue of Gödel's theorem for aesthetics would therefore be: There is no school of art which permits the production of all beauty and excludes the production of all ugliness'.

Prospective properties are beyond the reach of mere technique. They are outside the grasp of any mathematical Theory of Everything. That is why no non-poetic account of reality can be complete.

Summary

Theorem 100: This is the last theorem in the book.
(The proof is obvious.)

JOHN HORTON CONWAY[36]

Patterns are needed for conscious life to exist. Our descriptions of those patterns are what we call mathematics. Yet, we can extend those patterns far beyond the kaleidoscope provided by physical reality. We have taken a first glimpse at the subtlety that hides behind the patterned mask of mathematics. In the early years of this century, mathematicians like Hilbert set about banishing paradox and undecidability from the enterprise of mathematics. Despite good beginnings, the end was as no one imagined. Instead of arriving at a definition of mathematics that ensured that it was logically consistent and complete, within its own terms of reference, quite the opposite happened. Arithmetic could not prove itself to be consistent and complete. In this chapter we have begun to explore the ideas that went into that deduction, and to explore what physical examples of it might mean. Superficially, the idea that nature might be inconsistent in any sense seems inconceivable. But perhaps the possibility of time travel allows just such a possibility. We looked at the ideas of science, theology, and science fiction—with a curious contribution from investment banking as well—to evaluate whether time travel is really physically paradoxical.

The discovery of incompleteness and undecidability led to the discovery that there were further limitations upon our ability to get at the truths of mathematics that are decidable. Alan Turing first conceived of 'computers', defined by a series of step-by-step processes. He showed that these devices can establish only part of the collection of decidable truths.

The discoveries of Gödel and Turing have created a wave of modern interest in the consequences of their work for philosophy. These were, however, by no means the first 'impossibility' theorems to be proved by mathematicians. In the nineteenth century, the first proofs were given that certain geometrical constructions, like the trisecting of an angle with rules and compasses, are impossible. Strangely, these demonstrations did not capture the attention of non-mathematicians, or motivate wider consideration of the limits of mathematical reasoning about the physical world. Finally, we saw how the insights of Gödel and Turing allow us to isolate attributes that transcend the grasp of axioms and rules.

A new type of impossibility has emerged in this chapter, one that can be proved to exist, one that limits our most vigorous systems of reasoning, and one that threatens consequences for all our applications of reasoning to understand the Universe around us.

CHAPTER 8

Impossibility and us

Of the three principal sources of impossibility in politics—bureaucracies, factions, and elections—the single greatest source of impossibility is bureaucracy.

ADAM YARMOLINSKY[1]

Gödel's theorem and physics

The Kafkesque aspect of Gödel's work and character is expressed in his famous Incompleteness Theorem . . . Scientists are thus left in a position somewhat like Kafka in The Castle. *Endlessly, we hurry up and down corridors, meeting people, knocking on doors, conducting our investigations. But the ultimate success will never be ours. Nowhere in the castle of science is there a final exit to the absolute truth.*

RUDY RUCKER[2]

Gödel's monumental demonstration that systems of mathematics have limits gradually infiltrated the way in which philosophers and scientists viewed the world and our quest to understand it. Superficially, it appears that all human investigations of the Universe must be limited. Science is based on mathematics; mathematics cannot discover all truths; therefore science cannot discover all truths. This was how the argument went. Commentators with some religious apologetic in mind seized upon the limit to the power of human reason that Gödel implied. One of Gödel's contemporaries, and a student of Hilbert's, Hermann Weyl, described Gödel's discovery as exercising a 'constant drain on the enthusiasm' with which he pursued his scientific research. He believed that this underlying pessimism, so different from the rallying cry with which Hilbert had issued to mathematicians in 1900, was shared 'by other mathematicians who are not indifferent to what their scientific endeavours mean in the context of man's whole caring and knowing, suffering and creative existence in the world'. In more recent times, a frequent writer on theology and science, Stanley Jaki, believes that Gödel prevents us from gaining an understanding of the cosmos as a necessary truth,

Clearly then no scientific cosmology, which of necessity must be highly mathematical, can have its proof of consistency within itself as far as mathematics goes. In the absence of such consistency, all mathematical models, all theories of elementary particles, including the theory of quarks and gluons ... fall inherently short of being that theory which shows in virtue of its *a priori* truth that the world can only be what it is and nothing else. This is true even if the theory happened to account with perfect accuracy for all phenomena of the physical world known at a particular time.[3]

Jaki also sees Gödel's incompleteness theorem as a fundamental barrier to understanding of the Universe:

It seems on the strength of Gödel's theorem that the ultimate foundations of the bold symbolic constructions of mathematical physics will remain embedded forever in that deeper level of thinking characterized both by the wisdom and by the haziness of analogies and intuitions. For the speculative physicist this implies that there are limits to the precision of certainty, that even in the pure thinking of theoretical physics there is a boundary ... An integral part of this boundary is the scientist himself, as a thinker ...[4]

In the past decade, Gödel's insights have been further illuminated by casting them into the language of information and randomness in the manner pioneered by Greg Chaitin.[5] This has created a different way of viewing the implications for physics. Science is the search for compressions of strings of data into briefer encodings ('laws of Nature') which contain the same information. Any string of symbols which can be replaced by a formula or a rule that is shorter than the string itself will be called *compressible*. Any string that cannot be abbreviated in this way we call *incompressible*. We can always demonstrate that a given string is compressible by displaying the pattern that allows a compression of its information content to be made. But, strikingly, there is no way in which a general string of symbols can be proved to be incompressible. The pattern needed to abbreviate the string of symbols might be one of those truths which cannot be proved. Thus, you can never know whether your ultimate theory is the ultimate theory or not. There might always exist some deeper version of it: it might just be part of a larger theory.

These links between undecidability and randomness also allow us to forge further unexpected connections between Gödel and the efficiency of machines.[6] Undecidability will place limits on the efficiency of the machines of the far future. Suppose we take the example of a modern gas cooker. It is full of microprocessors, designed to sense the temperature inside the oven and implement instructions programmed into the control panel. The microprocessors store information temporarily until it is overwritten by new instructions or information. The more efficiently this information can be encoded and stored in the microprocessors, the more efficiently the cooker operates, because it minimizes the unneeded work carried out erasing and overwriting the instructions lodged in its memory.[7] But Chaitin's investigations show that Gödel's theorem is equivalent to the statement that we can never tell whether a program is the shortest one that will accomplish a given task. Hence, we can never find the most succinct program required to store the instructions for the operation of the cooker. As a result, the microprocessors we use will always overwrite more information than they need to: they will always possess some redundancy or inefficiency. In practice, this 'logical friction' produces a decrease in gas-cooker efficiency that is currently billions of times less than could be offset by simply cleaning it. Nonetheless, these considerations might one day prove important to the operation of delicate nanotechnological machines, and will be essential if we are to determine the ultimate capabilities of any technology.

Intriguingly, and just to show the important role that human psychology plays in assessing the significance of limits, some scientists, like Freeman Dyson, acknowledge that Gödel places limits on our ability to discover the truths of mathematics and science, but interpret this as ensuring that science will go on for ever. Dyson sees the incompleteness theorem as an insurance policy against the scientific enterprise, which he admires so much, coming to a self-satisfied end; for

> Gödel proved that the world of pure mathematics is inexhaustible; no finite set of axioms and rules of inference can ever encompass the whole of mathematics; given any set of axioms, we can find meaningful mathematical questions which the axioms leave unanswered. I hope that an analogous situation exists in the physical world. If my view of the future is correct, it means that the world of physics and astronomy is also inexhaustible; no matter how far we go into the future, there will always be new things happening, new information coming in, new worlds to explore, a constantly expanding domain of life, consciousness, and memory.

Thus, we see the optimistic and the pessimistic responses to Gödel. The optimists, like Dyson, see his result as a guarantor of the never-ending character of human investigation. They see scientific research as part of an essential part of the human spirit which, if it were completed, would have a disastrous effect upon us. Karl Popper had this in mind when he wrote that 'continued growth is essential to the rational and empirical character of scientific knowledge; that if science ceases to grow it must lose that character.' The pessimists, like Jaki, by contrast, interpret Gödel as establishing that the human mind cannot know all (perhaps not even most) of the secrets of Nature. They place more emphasis upon the possession and application of knowledge than on the process of acquiring it. The pessimist does not see the principal human benefit of science as arising from the quest for knowledge itself.

On reflection we should not be too surprised that the same state of affairs elicits such diametrically opposed responses. Many things in life create the same hiatus. It all depends whether you think your glass is half empty or half full. Gödel's own view was as unexpected as ever. He thought that intuition, by which we can 'see' truths of mathematics and science, was a tool that would one day be valued just as formally and reverently as logic itself:

> I don't see any reason why we should have less confidence in this kind of perception, i.e., in mathematical intuition, than in sense perception, which induces us to build up physical theories and to expect that future sense perceptions will agree with them and, moreover, to believe that a question not decidable now has meaning and may be decided in the future.[8]

Gödel was not minded to draw any strong conclusions for physics from his incompleteness theorems. He made no connections with the Uncertainty Principle of quantum mechanics, another great deduction that limited our ability to know, which was discovered by Heisenberg just a few years before Gödel made his discovery. In fact, Gödel was rather hostile to any consideration of quantum mechanics at all. Those who worked at the same Institute (no one really worked *with* him) believed that this was a result of his frequent discussions with Einstein, who, in the words of John Wheeler (who knew them both) 'brainwashed Gödel' into disbelieving quantum mechanics and the Uncertainty Principle. Greg Chaitin records this account of Wheeler's attempt to draw Gödel out on the question of whether there is a connection between Gödel incompleteness and Heisenberg Uncertainty:

> Well, one day I was at the Institute of Advanced Study, and I went to Gödel's office, and there was Gödel. It was winter and Gödel had an electric heater and had his legs wrapped in a blanket. I said 'Professor Gödel, what connection do you see between your incompleteness theorem and Heisenberg's uncertainty principle?' And Gödel got angry and threw me out of his office![9]

Does Gödel stymie physics?

> *Heavier-than-air flying machines are impossible.*
> LORD KELVIN (1895)

The argument that mathematics contains unprovable statements, physics is based on mathematics, and therefore physics will not be able to discover everything that is true, has been around for a long time. More sophisticated versions of it have been constructed which exploit the possibility of uncomputable mathematical operations being required to make predictions about observable quantities. From this vantage point the mathematical physicist Stephen Wolfram has conjectured that

> One may speculate that undecidability is common in all but the most trivial physical theories. Even simply formulated problems in theoretical physics may be found to be provably insoluble.[10]

Indeed, it is known that undecidability is the rule rather than the exception amongst the truths of arithmetic.[11]

With these worries in mind, let us look a little more closely at what Gödel's result might have to say about the course of physics. The situation is not so clear-cut as the commentators would have us believe. It is useful to lay out the precise assumptions that underlie Gödel's deduction of incompleteness. Gödel's theorem shows that if a formal system is (1) *finitely specified*, (2) *large enough to include arithmetic*, and (3) *consistent*, then it is *incomplete*.

Condition 1 means that there is a listable infinity of axioms. There must be a definite algorithmic procedure for listing them. We could not, for instance, choose our system to consist of all the true statements about arithmetic, because this collection cannot be finitely listed in this sense.

Condition 2 means that the formal system includes all the symbols and axioms used in arithmetic. The symbols are 0, ('zero'), S, ('successor of'), $+$, $\times$, and $=$. Hence, the number two is the successor of the successor of zero, written as the term SS0, and 'two and plus two equals four' is expressed as $SS0 + SS0 = SSSS0$.

The structure of arithmetic plays a central role in the proof of Gödel's theorem. Special properties of numbers, like their primeness and the fact that any number can be expressed in only one way as the product of the prime numbers that divide it, were used by Gödel to establish the vital correspondence between statements of mathematics and statements about mathematics. In this way, linguistic paradoxes like that of the 'liar' could be embedded, like Trojan horses, within the structure of mathematics itself. Only logical systems which are rich enough to include arithmetic allow this incestuous encoding of statements about themselves to be made within their own language.

Again, it is instructive to see how these requirements might fail to be met. If we picked a theory that consisted of references to (and relations between) only the first ten numbers (0, 1, 2, 3, 4, 5, 6, 7, 8, 9), then Condition 2 fails and such a mini-arithmetic is complete. Arithmetic makes statements about individual numbers, or terms (like SS0, above). If a system does not have individual terms like this but, like Euclidean geometry, makes statements only about points, circles, and lines in general, then it cannot satisfy Condition 2. Accordingly, as Alfred Tarski first showed, Euclidean geometry is complete. There is nothing magical about the flat, Euclidean nature of the geometry either: the non-Euclidean geometries on curved surfaces are also complete. Similarly, if we had a logical theory dealing with numbers that used only the concept of 'greater than' without referring to any specific numbers, then it would be complete: we can determine the truth or falsity of any statement about numbers involving the 'greater than' relationship.

Another example of a system that is smaller than arithmetic is arithmetic without the multiplication, ×, operation. This is called Presburger arithmetic (the full arithmetic is called Peano arithmetic after the mathematician who first expressed it axiomatically, in 1889). At first this sounds strange. In our everyday encounters with multiplication it is nothing more than a shorthand way of doing addition (for example, $2+2+2+2+2+2 = 2×6$), but in the full logical system of arithmetic, in the presence of logical quantifiers like 'there exists' or 'for any', multiplication permits constructions which are not merely equivalent to a succession of additions.

Gödel showed, as part of his doctoral thesis work, that Presburger arithmetic is complete: all statements about the addition of natural numbers can be proved or disproved; all truths can be reached from the axioms.[12] Similarly, if we create another truncated version of arithmetic which does not have addition but retains multiplication, this is also complete. It is only when addition and multiplication are simultaneously present that incompleteness emerges. Extending the system further by adding extra operations like exponentiation to the repertoire of basic operations makes no difference. Incompleteness remains, but no intrinsically new form of it is found. Arithmetic is the watershed in complexity.

The use of Gödel to place limits on what a mathematical theory of physics (or anything else) can ultimately tell us seems a fairly straightforward consequence. But as one looks more carefully into the question, things are not quite so simple. Suppose, for the moment, that all the conditions required for Gödel's theorem to hold are in place. What would incompleteness look like in practice? We are familiar with the situation of having a physical theory which makes accurate predictions about a wide range of observed phenonema: we might call it 'the standard model'. One day, we may be surprised by an observation about which it has nothing to say. It cannot be accommodated within its framework. Examples are provided by some so-called grand unified theories in particle physics. Some early editions of these theories had the property that all neutrinos must have zero mass. Now if a neutrino is observed to have a non-zero mass (as everyone believes it will have, and some experiments have even claimed to have measured), then we know that the new situation cannot be accommodated within our original theory. What do we do? We have encountered a certain sort of incompleteness, but we respond to it by extending or modifying the theory to include the new possibilities. Thus, in practice incompleteness looks very much like inadequacy in a theory.

In the case of arithmetic, if some statement about arithmetic is known to be undecidable (there are known statements of this sort; it means that both their truth and falsity are consistent with the axioms of arithmetic), then we have two ways of extending the structure. We can create two new arithmetics: one which adds the undecidable statement as an extra axiom, the other which adds its negation as a new axiom. Of course, the new arithmetics will still be incomplete, but they can always be extended to accommodate any incompleteness. Thus, in practice, a physical theory can always be enlarged by adding new principles which force all the undecidability into the part of the mathematical realm which has no physical manifestation. Incompleteness would then always be very hard, if not impossible, to distinguish from incorrectness or inadequacy.

An interesting example of this dilemma is provided by the history of mathematics. During the sixteenth century, mathematicians started to explore what happened when they added together infinite lists of numbers. If the quantities in the list get larger then the sum will 'diverge', that is, as the number of terms approaches infinity so does the sum. An example is the sum

$$1 + 2 + 3 + 4 + 5 + ... = \text{infinity}.$$

However, if the individual terms get smaller and smaller sufficiently rapidly,[13] then the sum of an infinite number of terms can get closer and closer to a finite limiting value which we shall call the sum of the series; for example

$$1 + 1/9 + 1/25 + 1/36 + 1/49 + ... = \pi^2/8 = 1.2337005....$$

This left mathematicians to worry about a most peculiar type of unending sum,

$$1 - 1 + 1 - 1 + 1 - 1 + 1 - ..... = ?????$$

If you divide up the series into pairs of terms it looks like $(1-1)+(1-1)+....$ and so on. This is just $0+0+0+... = 0$ and the sum is zero. But think of the series as $1-\{(1-1)+(1-1)+(1-1)+...\}$ and it looks like $1 - \{0\} = 1$. We seem to have proved that $0 = 1$.

Mathematicians had a variety of choices when faced with ambiguous sums like this. They could reject infinities in mathematics and deal only with finite sums of numbers, or, as Cauchy showed in the early nineteenth century, the sum of a series like the last one must be defined by specifying more closely what is meant by its sum. The limiting value of the sum must be specified together with the procedure used to calculate it. The contradiction $0 = 1$ arises only when one omits to specify the procedure used to work out the sum. In the two cases it is different and so the two answers are not the same. Thus, here we see a simple example of how a limit is sidestepped by enlarging a concept which seems to create limitations. Divergent series can be dealt with consistently so long as the concept of a sum for a series is suitably extended.[14]

Another consideration about Gödel's theorem is that the physical world makes use only of the decidable part of mathematics. We know that mathematics is an infinite sea of possible structures. Only some of those structures and patterns appear to find existence and application in the physical world. It may be that they are all from the subset of decidable truths. The hierarchy is illustrated in Figs. 8.1 and 8.2.

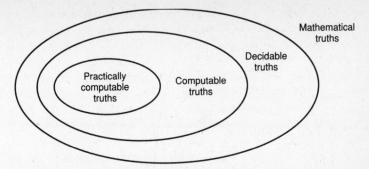

Fig. 8.1 The 'universe' of all mathematical truths contains subsets which contain all decidable truths, all computable truths, and all practically computable truths (that is, all computations that can be completed in, say, the age of the Universe, about 15 billion years; see Fig. 8.2).

It is also possible that the conditions required to prove Gödel's incompleteness do not apply to physical theories. Condition 1 requires the axioms of the theory to be listable. It might be that the laws of physics are not listable in this predictable sense. This would be a radical departure from the situation that we think exists, where the number of fundamental laws is believed to be not just listable, but finite (and very small). But it is always possible that we are just scratching the surface of a bottomless tower of laws, only the top of which has significant effects upon our experience. However, if there were an unlistable infinite of physical laws then we would face a more formidable problem than that of incompleteness.

In fact, in 1940, Gerhard Gentzen, one of Hilbert's young students who lost his life in the Second World War soon afterwards, showed that it was possible to circumvent Gödel's conclusions and deduce all the truths of arithmetic if a procedure of transfinite induction is included. Again, the operations of Nature might include such a non-finite system of axioms. We are inclined to think of incompleteness as something undesirable because it implies that we shall not be able to 'do' something. But we could turn the situation on its head and conclude that Nature is consistent and complete but cannot be captured by a finite set of axioms. There is something aesthetically satisfying about this superhuman complexion to things.

An equally interesting issue is that of finiteness. It may be that the universe of physical possibilities is finite, although astronomically large. However, no matter how large the number of primitive quantities to which the laws refer, so long as they are finite the resulting system of interrelationships will be complete. We should stress that although we habitually assume that there is a continuum of points of space and time, this is just an assumption that is very convenient for the use of simple mathematics. There is no deep reason to believe that space and

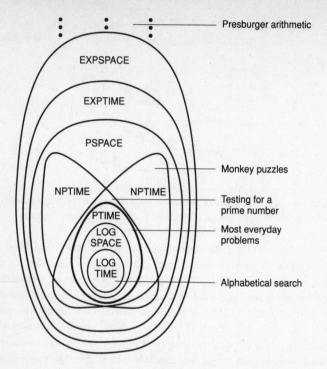

Fig. 8.2 Some finer detail about the practically computable truths. There is a hierarchy of complexity according to the computer processing time (TIME) and memory space (SPACE) required to solve problems. For the simplest problems (LOGSPACE and LOGTIME) the computational requirements grow only logarithmically as the number of inputs increases. The next level of polynomially complex problems (PTIME and PSPACE) are the most common. The realm of computational intractability can be divided into problems that require exponentially increasing computational resources (EXPTIME and EXPSPACE) as the number of inputs increases, and those like Presburger arithmetic that possess double (or higher) exponential complexity. Some sample problems, discussed in Chapter 4, are located in this scheme.[61]

time are continuous, rather than discrete, at their most fundamental micro-scopic level; in fact, there are some theories of quantum gravity that assume that they are not. Quantum theory has introduced discreteness and finiteness in a number of places where once we believed in a continuum of possibilities. Curiously, if we give up this continuity, so that there is not necessarily another point in between any two sufficiently close points you care to choose, space-time structure becomes vastly more complicated. Many more complicated things can happen. This question of finiteness might also be bound up with the question of whether the Universe is finite in volume and whether the number of elementary particles (or whatever the most elementary entities might be) of

Nature are finite or infinite in number. Thus there might exist only a finite number of terms to which the ultimate logical theory of the physical world applies. Hence, it would be complete.

An interesting possibility with regard to the application of Gödel's results to the laws of physics is that Condition 2 of the incompleteness theorem might not be met. How could this be? Although we seem to make wide use of arithmetic, and much larger mathematical structures, when we carry out scientific investigations of the laws of Nature, this does not mean that the inner logic of the physical Universe needs to employ such a large structure. It is undoubtedly convenient for us to use large mathematical structures, together with concepts like infinity, but this may be an anthropomorphism. The deep structure of the Universe may be rooted in a much simpler logic than that of full arithmetic, and hence be complete. All this would require would be for the underlying structure to contain either addition or multiplication but not both. Recall that all the sums that you have ever done have used multiplication simply as a shorthand for addition. They would be possible in Presburger arithmetic as well. Alternatively, a basic structure of reality that made use of simple relationships of a geometrical variety, or which derived from 'greater than' or 'less than' relationships, or subtle combinations of them all could also remain complete.[15] The fact that Einstein's theory of general relativity replaces many physical notions like force and weight by *geometrical* distortions in the fabric of space-time may well hold some clue about what is possible here.

The laws of physics might be fully expressible in terms of a mathematical system that is complete, but in practice we would always be far more concerned with making sure that we had got the *correct* system rather than a complete system.

There is another important aspect of the situation to be kept in view. Even if a logical system is complete, it always contains unprovable 'truths'. These are the axioms which are chosen to define the system. And after they are chosen, all the logical system can do is deduce conclusions from them. In simple logical systems, like Peano arithmetic, the axioms seem reasonably obvious because we are thinking backwards—formalizing something that we have been doing intuitively for thousands of years. When we look at a subject like physics, there are parallels and differences. The axioms, or laws, of physics are the prime target of physics research. They are by no means intuitively obvious, because they govern regimes that can lie far outside our experience. The outcomes of those laws are unpredictable in certain circumstances because they involve symmetry-breakings. Trying to deduce the laws from the outcomes is not something that we can ever do uniquely and completely by means of a computer program.

Thus, we detect a completely different emphasis in the study of formal systems and in physical science. In mathematics and logic, we start by defining a system of axioms and laws of deduction. We might then try to show that the system is complete or incomplete, and deduce as many theorems as we can from the axioms. In science, we are not at liberty to pick any logical system of laws that we choose. We are trying to find the system of laws and axioms (assuming there is one—or more than one perhaps) that will give rise to the outcomes that we see. As we stressed earlier, it is always possible to find a system of laws which will give rise to any set of observed outcomes. But it is the very set of unprovable statements that the logicians and the mathematicians ignore—the axioms and laws of deduction—that the scientist is most interested in discovering rather than simply assuming. The only hope of proceeding as the logicians do would be if for some reason there were only one possible set of axioms or laws of physics. So far, this remains a possibility;[16] but even if it were the case, we would not be able to prove it.

Specific examples have been given of physical problems which are undecidable. As one might expect from what has just been said, they do not involve an inability to determine something fundamental about the nature of the laws of physics or the most elementary particles of matter. Rather, they involve an inability to perform some specific mathematical calculation, which inhibits our ability to determine the course of events in a well-defined physical problem. However, although the problem may be mathematically well defined, this does not mean that it is possible to create the precise conditions required for the undecidability to exist.

An interesting series of examples of this sort has been created by the Brazilian mathematicians Francisco Doria and Newton da Costa.[17] Responding to a challenge problem posed by the Russian mathematician Vladimir Arnold, they investigated whether it was possible to have a general mathematical criterion which would decide whether or not any equilibrium was stable. A stable equilibrium is a situation like a ball sitting in the bottom of a basin—displace it slightly and it returns to the bottom; an unstable equilibrium is like a needle balanced vertically—displace it slightly and it moved away from the vertical[18]. The two situations are pictured in Fig. 8.3.

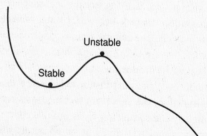

Unstable

Stable

Fig. 8.3 Stable and unstable equilibria. A ball at rest at two possible locations.

When the equilibrium is of a simple nature this problem is very elementary; first-year science students learn about it. But when the equilibrium exists in the face of more complicated couplings between the different competing influences, the problem soon becomes more complicated. So long as there are only a few competing influences, the stability of the equilibrium can still be decided by inspecting the equations that govern the situation. Arnold's challenge was to discover an algorithm which tells us whether this can always be done, no matter how many competing influences there are, and no matter how complex their interrelationships. By 'discover' he meant find a formula into which you can feed the equations which govern the equilibrium together with your definition of stability, and out of which will pop the answer 'stable' or 'unstable'.

Strikingly, da Costa and Doria discovered that there can exist no such algorithm. There exist equilibria characterized by special solutions of mathematical equations whose stability is undecidable. In order for this undecidability to have an impact on problems of real interest in mathematical physics, the equilibria have to involve the interplay of very large numbers of different forces. While such equilibria cannot be ruled out, they have not yet arisen in real physical problems. Da Costa and Doria went on to identify similar problems where the answer to a simple question, such as 'will the orbit of a particle become chaotic?', is Gödel-undecidable. Others have also tried to identify formally undecidable problems. Geroch and Hartle have discussed problems in quantum gravity that predict the values of potentially observable quantities as a sum of terms whose listing is known to be a Turing-uncomputable operation.[19] Pour-El and Richards showed that very simple differential equations which are widely used in physics, like the wave equation, can have uncomputable outcomes when the initial data are not very smooth.[20] This lack of smoothness gives rise to what mathematicians call an 'ill-posed' problem. It is this feature that gives rise to the uncomputability. However, Traub and Wozniakowski have shown that every ill-posed problem is well-posed on the average under rather general conditions.[21] Wolfram gives examples of intractability and undecidability arising in condensed-matter physics.[22]

The study of Einstein's general theory of relativity also produces an undecidable problem if the mathematical quantities involved are unrestricted.[23] When one finds an exact solution of Eintein's equations it is always necessary to discover whether it is just another, known solution that is written in a different form. One can usually investigate this by hand, but for complicated solutions computers can help. For this purpose we require computers programmed for algebraic manipulations. They can check various quantities to discover if a given solution is equivalent to one already sitting in its memory bank of known solutions. In the practical cases encountered so far, this checking procedure comes up with a definite result after a small number of steps. But in general the comparison is an undecidable process equivalent to another famous undecidable problem of pure mathematics, 'the word problem' of group theory.

The conclusion we should draw from this discussion is that it is by no means obvious that Gödel places any straightforward limit upon the overall scope of physics to understand the nature of the Universe just because physics makes use of mathematics. The mathematics that Nature makes use of may be smaller and simpler than is needed for incompleteness and undecidability to rear their heads. Yet, within science, it is the smaller individual problems that are at the mercy of computational intractability and undecidability.

Gödel, logic, and the human mind

> *I believe that people would be alive today if there were a death penalty.*
>
> NANCY REAGAN

One persistent use of Gödel's results has been to argue that in some way the human mind is superior to computing machines. Paradoxically, it is the very fallibility of the human mind to which some appeal in order to argue for its superiority over machines. Because the machine blindly follows the laws of logic that are programmed into it, it is under the spell of Gödel's theorem, and cannot ascertain the truth or falsity of all statements in its language. The human mind on the other hand, so the argument goes, is not a slave of deductive arguments. It can use intuition, guesswork, induction, and all other means of non-deductive reasoning to get at the truth.

We hear Jaki claim that

> Gödel's theorem casts light on the immense superiority of the human brain over such of its products as the most advanced forms of computers[24]

or Nagel and Newman offer their opinion that

> Gödel's conclusions bear on the question whether a calculating machine can be constructed that would match the human brain in mathematical intelligence . . . as Gödel showed . . . there are innumerable problems in elementary number theory that fall outside the scope of a fixed axiomatic method, and that such machines are incapable of answering, however intricate and ingenious their built-in mechanisms may be and however rapid their operations . . . the brain appears to embody a structure of rules of operation which is far more powerful than the structure of currently conceived artificial machines . . . the resources of the human intellect have not been, and cannot be, fully formalised, and that new principles of demonstration forever await invention and discovery.[25]

Others have disagreed strongly with these conclusions. Here is the philosopher Michael Scriven on Nagel and Newman.

Nagel and Newman are struck by the fact that whatever axioms and rules of inference one might give a computer, there would apparently be mathematical truths which it would never 'reach' from these axioms by the use of these rules. This is true, but their assumption that we could suppose ourselves to have given the machine an adequate idea of mathematical truth when we give it the axioms and rules of inference is not true ... The Gödel theorem is no more an obstacle to a computer than to ourselves ... But just as we can recognise the truth of the unprovable formula by comparing what it says with what we know to be the case, so can a computer do the same.[26]

The most famous argument for the superiority of human reasoning when pitted against computers was the paper entitled 'Minds, machines, and Gödel', by the Oxford philosopher John Lucas, who argued that

a conscious being can deal with Gödelian questions in a way in which a machine cannot, because a conscious being can both consider itself and its performance and yet not be other than that which did the performance. A machine can be made to in a manner of speaking to 'consider' its performance, but it cannot take this 'into account' without thereby becoming a different machine, namely the old machine with a 'new part' added. But ... a conscious mind ... can reflect upon itself ... and no extra part is required.[27]

This argument has attracted a host of critics from the cognitive sciences. A range can be found in the writings of Douglas Hofstadter.[28] Lucas's style of argument has been developed by Rudy Rucker and by Roger Penrose.[29] Rucker considers an ultimate artificial machine intelligence, which he calls the Universal Truth Machine (UTM). He shows that Gödel can construct a truth which the UTM can never utter. Gödel's sentence is therefore

a specific mathematical problem that we know the answer to, even though UTM does not! So UTM does not, and cannot, embody a best and final theory of mathematics.[30]

Penrose has reiterated this argument and used it as a springboard to argue for specific non-algorithmic processes operating in the brain. Again, there have been a host of criticisms of this sweeping argument, some of which are dealt with by Penrose.[31] However, the most interesting response to all these appeals to Gödel sentences which prove their own provability as examples of the superiority of human intuition over machine 'intelligence' arose during a debate between John Lucas, the cognitive scientist Christopher Longuet-Higgins, and the philosopher Anthony Kenny, on 'The Nature of the Mind', which formed part of the 1970 Gifford Lectures at the University of Edinburgh.[32] It highlights the symmetrical relationship between humans and machines in respect of each being able to assert what the other cannot. Each person and each machine has a statement that others cannot logically assert, but this does not endow any who make their assertion with special abilities. Kenny takes up the discussion:

> You remember that John Lucas argued that minds were not machines because, given any machine working algorithmically, we could produce something which would be like a Gödelian formula . . . we could present it with a formula which we could see to be true, but the machine couldn't prove to be true . . . one of his critics . . . said, 'Take this sentence: "John Lucas cannot consistently make this judgement" . . . Clearly any other human being except John Lucas can see this is true, without inconsistency. But clearly John can't make this judgement without inconsistency, therefore that shows that we all have a property which he doesn't have, which makes us as much superior to him as we all are to computers . . .'

A similar argument was brought to bear on the arguments of Penrose, in greater detail, by the computer scientist John McCarthy when reviewing *The Emperor's New Mind*.[33] In all these debates, a single assumption is always lurking beneath the surface. It is the assumption that the workings of the brain are infallible, when viewed as logical processors. There is really no reason to believe this (and many reasons not to!). The brain is a staging point in an ongoing evolutionary process. The mind was not evolved for the 'purpose' of doing mathematics. Like most evolutionary products it does not need to be perfect, merely better than previous editions, and sufficiently good to endow a selective advantage. If we admit that the mind is fallible, then the assessment of Gödelian sentences is beside the point. We would need to conclude that the mind was ultimately inconsistent rather than incomplete. As a result there is nothing more to be said with regard to its parity with algorithmic machines.

The problem of free will

We have to believe in free will. We've got no choice.

ISAAC SINGER[34]

The application of Gödel's style of argumentation to questions of complete self-knowledge, free will, and determinism was first made by the late Karl Popper, in a pair of articles written for the first issue of the *British Journal for the Philosophy of Science* in 1950.[35] Popper showed that a deterministic computing machine could not produce a prediction of its own future state which would remain valid if it was embodied within itself, because the process of embodiment would inevitably render it out of date. Physicists were familiar with the heuristic picture of Heisenberg's Uncertainty Principle making perfect measurement an impossibility, because the very act of measurement would perturb the system by a relatively larger and larger amount as the dimension being probed got smaller and smaller. Popper used the logical equivalent of this perturbation, which is a simple consequence of the earlier arguments of Gödel and Turing, to limit the ability of a computer to understand and predict its behaviour completely: complete self-description is logically impossible. The dilemma is not unlike that of the fictional Tristram Shandy, who found his autobiography unable to keep up with his rate of living, for

> In order to predict oneself completely, one has to predict oneself predicting oneself completely, and then one has to further predict oneself predicting oneself predicting oneself completely. The infinite regress is clear.[36]

This argument was taken up and applied more specifically to theological and philosophical questions by the British cognitive scientist, Donald Mackay. Mackay was a frequent writer on issues of common interest to religion and science. His style was spare and logical, and his approach gave glimpses of his underlying Calvinist background. He had a long-standing interest in matters of free will and determinism, and strove to use the arguments of Gödel and Popper to clarify the confused discussion which he detected in most discussions of determinism, predestination, and free will. His arguments, although logically precise and rigorous, were quite straightforward and appeared in many magazines directed at general readers, notably first in two issues of *The Listener*, the weekly magazine of the BBC, in May 1957.

Mackay asks us to consider a world which is totally deterministic (forget about things like quantum-mechanical uncertainty and the finite sensitivity of measuring devices for the moment); all phenomena, even personal decisions and opinions, are supposed to be determined completely in advance by a system of rigid laws of Nature. Laplace's vision is realized. Now we ask, would it be possible, even in principle, to predict someone else's behaviour completely in this world?

At first sight, you might think that it would. But look more closely. Consider a person who is asked to choose between soup or salad for lunch. If we introduce a brain scientist who not only knows the complete state of this person's brain, but that of the entire universe as well at present, we could ask whether this scientist can infallibly announce what the choice of lunch will be. The answer is 'no'. The subject can always be stubborn, and adopt a strategy that says, 'If you say that I will choose soup, then I will choose salad, and vice versa'. Under these conditions it is logically impossible for the scientist to predict infallibly what the person will choose if the scientist makes his prediction known.

This does not mean that it is impossible for the scientist to *know* infallibly what the person's choice will be. So long as he keeps this knowledge to himself, his deterministic theory of the diner's thoughts and actions can continue to be infallible. He could tell other people. He could even write the prediction down on a piece of paper and show it to the diner after he had chosen his lunch. In both cases, he could have predicted correctly, but would not have exercised any constraint upon the diner's free choice that the diner knew of. It is only when he decides to make it known to the diner that the scales are tipped against him and the diner can always falsify his prediction if he chooses to. As the prediction is made known, it cannot be unconditionally binding on the person whose actions it predicts. That person may always act so as to falsify the prediction. He doesn't have to, but he may; you cannot be sure.

Let us unpack the argument a little further. Suppose we are in possession of a complete theory for predicting your next action, if we know your current brain state. We demonstrate how good we are at doing this by showing our predictions to other people, all of whom confirm that you act precisely as predicted. Suppose your brain is in state 1 and we predict you will act as $P(1)$. Would you be correct to believe the prediction $P(1)$ if it were shown to you?

First, we must consider the effect on your brain state of believing the prediction $P(1)$. If believing the prediction changed the state of your brain to state 2, then the act of believing the prediction $P(1)$ would put your brain into a different state from that on which the prediction was based. The new brain state 2 would give rise to a new prediction $P(2)$. The key question is whether we can build into our predictions the effects of making the prediction $P(1)$ known to you, so that we could make the prediction $P(2)$. But, if that were done, we could not claim that $P(2)$ is what you would be correct to believe, because it is brain state 2 that leads to prediction $P(2)$, and if you believed $P(2)$ this would again change your brain state from state 2 to some new state 3, say, and $P(2)$ would *not* be a correct prediction of the action that follows from that state. The accuracy of any prediction we can make of your behaviour is conditional on your not believing it.

This is an interesting state of affairs. Usually, we think of something that is 'true' as being true for everyone. Here, this universality does not exist. The correlation between brain states and knowledge creates a logical indeterminacy about the future: there is a distinction between something being predictable by others and inevitable for oneself.

Mackay's aim here was to show that a deterministic model of brain action would not render untenable the belief that individuals enjoy freedom of choice (under normal circumstances). He makes no appeal to quantum uncertainties or non-computability. He also makes the strongest possible assumption about the brain's encoding of a person's thoughts and feelings: that everything they see, hear, feel, believe, etc. is completely and uniquely encoded in their physical brain state. Thus, a change of belief about something (that is, a change of mind) would be representable by a specific transformation from one brain state to another.

Explaining what he means by 'freedom', Mackay writes:

> by calling a man 'free', (a) we might mean that this action was *unpredictable by anyone*. This I would call freedom of caprice; or (b) we may mean that the outcome of his decision is *up to him*, in the sense that unless he makes the decision it will not be made, that he is in a position to make it, and that no fully-determinate specification of the outcome already exists, which he would be correct to accept as inevitable, and would be unable to falsify, if only he knew it.[37]

Mackay applies this to the question of divine foreknowledge to argue

> that divine foreknowledge is not something that *we* would be correct to believe if only we knew it—since for us (unlike God) this would involve a contradiction.[38]

From this, he goes on to conclude that physical determinism (of neural processing) does not imply 'metaphysical determinism (denying the reality of human freedom and responsibility)'. Moreover, what many have traditionally regarded as the theological doctrine of predestination is logically impossible. Past disputes have argued over a serious misunderstanding of the logic of the situation:

> This may sound strange to those of us who have been accustomed to suppose that the doctrine of divine predestination meant just this—that there already exists now a description of us and our future, including the choices we have not yet made, which is binding upon us, if only we knew it, because it is known to God. But I hope that it is now clear that we should do God no honour by such a claim; for we should merely be inviting ourselves to imagine him in a logical self-contradiction. At this moment, we are unaware of any such description; so if it existed it would have to describe us as *not believing it*. But in that case we would be in error to believe it, for our believing it would falsify it! On the other hand, it would be of no use to alter the description so that it describes us as *believing it*; for in that case it is at the moment false, and therefore, although it would become correct if we believed it, we are not in error to *dis*believe it! Thus the divine foreknowledge of our future, oddly enough, has no unconditional logical claim upon us, unknown to us.
>
> This, I believe, demonstrates a fallacy underlying both the theological dispute between Arminianism and Calvinism, and the philosophical dispute between physical or psychological determinism . . . and libertarianism in relation to man's responsibility . . . even God's sovereignty over every twist and turn of our drama does not contradict . . . our belief that we are free, in the sense that no determining specification already exists which if only we knew it we should be correct to believe and in error to disbelieve, whether we liked it or not.[39]

These arguments have a clear and simple message for any type of predictive and explanatory study. There are unpredictable aspects of completely deterministic phenomena.[40]

There is a further dilemma that can be created from the arguments of Popper and Mackay. For Mackay envisages a Superbeing doing the predicting, and the subject of the Superbeing's predictions as being two different 'minds'. But what if they were one and the same? Suppose I know so much about the workings of the brain and the outside Universe that I can compute what I will choose to eat for dinner? Suppose further that I am rather perverse, and so decide that I will deliberately choose not to eat whatever it is that my calculations predict that I will. I have thus succeeded in making it logically impossible for me to predict what I will choose. Yet, if I had decided to be sensible I could have decided deliberately to choose to eat whatever it is that my calculations predict that I will choose to eat. In that case, I am successfully able to predict my future actions— but only if I choose to do so. Paradoxically, it seems that it is in my power to decide whether I can predict my future or not.

Let us look at what sort of dilemma this creates for our Superbeing. If he stubbornly chooses to act contrary to what his predictions say he will do, he cannot predict the future, even if the Universe is completely deterministic. He cannot therefore know the whole structure of the Universe. Omniscience is logically impossible for him, if he wants to be contrary. But if he doesn't want to be contrary, then he can be omniscient. No being can predict what he will do if he will not do what he predicts he will do!

The reaction game

You can only predict things after they have happened.

EUGÈNE IONESCO[41]

It has been said that economic forecasting is not like weather forecasting: economic forecasting can change the economy but weather forecasting can't change the weather. An activity like economic forecasting displays the inevitable dependence of what is being forecast on the forecasting process that was displayed by Mackay's consideration of human choices. However, despite this self-evident problem, it is striking to find the Nobel prize-winning economist Herbert Simon making the erroneous claim that it *is* possible to make predictions of elections that are automatically adjusted to take voter reaction into account. Political scientists have dubbed the problem 'The Reaction Paradox', but seem completely unaware of the work of Popper and Mackay on the problem.[42] In fact, Simon goes so far as to claim that his result 'refutes, therefore, that it would be *logically impossible* to make an accurate prediction [of public predictions] . . .'[43]

In a paper written in 1954, entitled 'Bandwagon and underdog effects in election predictions',[44] Simon claims to establish the possibility of making correct predictions, *even when the predictions are made known to the voters*, concluding, in contradiction to Popper and Mackay (of whose work he is unaware), that 'This proof refutes allegations commonly made about the impossibility, in principle, of correct prediction of social behaviour.'

In fact, Simon's proof is wrong. It makes illegal use of a theorem of mathematics called the Brouwer Fixed Point theorem.[45] There would need to be an infinitely large electorate and a continuum of predictions and responses in order for this theorem to be applicable. Remarkably, this false 'theorem' seems to occupy quite a prominent place in the literature of political science. One hopes that no election strategists are counting too strongly upon it. 'Practical politics', as Henry Brooks Adams once said, 'consists in ignoring facts'—but that's a risky way to operate.

In fact, Karl Aubert has shown that if the reaction problem is analysed correctly, with due regard for the fact that there are just a finite number of distinctive reactions to a pre-election forecast, and a *finite* number of outcomes to the election (rather than an infinite continuum), then we can determine the *probability* that an election forecast is accurate. If the election has n possible outcomes, and we assume that every possible reaction to a forecast is equally likely (which may not, of course, be the case), then the probability of a correct forecast for any n is given by the simple formula[46]

$$\text{Probability}(n) = 1 - (1 - 1/n)^n$$

When there is only one possible outcome, we have Probability($n=1$) = 1, and we can be 100 per cent of guessing right. With two possible outcomes our chance is 75 per cent; with three it falls to 70 per cent, and the value of Probability(n) gets steadily smaller as n increases, getting closer and closer to the value 0.63. This is surprising. There is a 63 per cent chance of predicting correctly as the number of outcomes gets very large. It is better than evens (i.e. 50 per cent), but it can never be 100 per cent.

There have been recent investigations which establish stronger, more wide-ranging, theorems about the impossibility of predicting the future.[47] It can be shown that it is impossible to build a computer which can correctly predict its future state before that state actually occurs. This impossibility exists even when one tries to predict the future states of finite systems that are non-chaotic (that is, they do not exhibit sensitive dependence on an imprecisely known initial state), and are assumed to manifest no quantum-mechanical uncertainty; this is true even if the computer is infinitely fast and is more powerful than Turing's idealized machine. This result is akin to a physical analogue of Gödel's theorem: it tells us that we cannot process information faster than the Universe does.

Mathematics that comes alive

What is it that puts fire in the equations and makes them come alive?

JOHN A. WHEELER

So far, we have regarded mathematics as something quite distinctive from the world which scientists study. It is a collection of all possible patterns from which one must choose a candidate to describe the way in which some aspect of Nature behaves. But there is a more unusual way of thinking, which I introduced in one of my earlier books, *Pi in the Sky*, about the nature of mathematics.

In physics and cosmology, we are used to envisaging a collection of all possible worlds and then asking the question: how small is the subset of all possibilities which permits the evolution of organized complexity which is complex enough to be called 'alive'? This set of life-supporting possibilities appears to be very small, in the sense that if many of the defining characteristics of the observable universe and the laws and constants of physics were slightly changed, then observers could not exist. If so, this discovery might be telling us something profound about the origin of the structure of the Universe and its defining characteristics.

Let us apply this style of thought to mathematics. In order to do so, we have to make a radical change of perspective. We have met two concepts of what is meant by 'existence'. For mathematicians of a formalistic persuasion, like Hilbert, existence meant nothing more than logical consistency: anything free from logical contradiction could exist in the mathematicians' world.[48] For the scientist, existence means that we observe it in the Universe. It must be physically real. Scientists usually assume that this must mean that physical reality is much smaller than the mathematicians' universe of logically consistent possibilities. But what if they were really one and the same?[49] This is a possibility which I explored in my book *Pi in the Sky*. We should imagine the collection of all possible mathematical systems, defined by all possible systems of axioms and rules of deduction. Now ask the question 'how complex does a mathematical structure need to be in order to permit the characterization of conscious observers?' If Penrose were correct about the link between consciousness and Gödel sentences, then we would require incompleteness in order to have consciousness in the formalism. But, as we have seen earlier in this chapter, this means that the formalism must be complex enough to contain arithmetic: geometry will not do. This inquiry can be pushed further, to see which aspects of logic, counting, and other basic mathematical patterns, might be necessary for the representation of organized complexity sufficiently exotic to describe life. If one adopts a thoroughgoing Platonic position (as did Gödel himself[50]) this view becomes very natural because the formalism is a representation of some another reality.

A stranger sort of impossibility

Elections are won by men and women chiefly because most people vote against
somebody rather than for somebody.

FRANKLIN P. ADAMS[51]

Social scientists and politicians have long been interested in the subtleties of voting. Today, voting is not confined to human electorates and ice-skating competitions; advanced technological systems, like space missions, are often under the control of a number of computers (an odd number!) which 'vote', on the basis of the data analysis they have performed, whether or not the launch takes place. If two vote 'abort' and one votes 'launch', the mission is aborted. Stranger still, there are serious theories of the workings of the human mind that picture it as a multi-levelled system of separate influences which interact rather like a society, each 'voting' for a particular course of action. Somehow a choice is finally made. This 'society of mind' picture pioneered by Marvin Minsky certainly strokes a resonant chord with our feelings of being in 'two minds', or of indecisiveness in the face of complex alternatives. Thus we might envisage that any form of natural complexity that is sufficient to produce self-reference, or allow conscious choices, will share any limitations that voting procedures might share.

In this chapter we have so far been looking at some of the ways in which the overall structure of logical systems can create impossibilities. We end by showing how it is possible to create collective impossibility by the addition of a number of perfectly rational individual choices.

Imagine that you are listening to the debate of the government's Star Chamber—an inner council of three who must take a far-reaching decision about the future of the country. You have three options in front of you: (1) have only a National Health Service; (2) have only a private medical insurance scheme; (3) have a mixture of the two systems.

The three members of the Star Chamber are A, B, and C. They each vote in order of their preferences: thus, A prefers policy 1 to policy 2 to policy 3; B prefers policy 2 to policy 3 to policy 1; C prefers policy 3 to policy 1 to policy 2. The civil servants carefully note down these preferences, and add up the voting scores. They see that policy 1 is preferred to policy 2 by a clear majority of two votes to one, and that policy 2 is preferred to policy 3 by two votes to one. 'That's it then,' announces A, 'we have only the NHS from now on; splendid!' 'Hang on a minute, there's something odd here,' Sir Humphrey interjects, 'policy 3 is preferred to policy 1 by two votes to one; 1 beats 2, and 2 beats 3, but 3 beats 1. what is going on, Minister?'

This little example is extremely worrying. It was first identified by a French mathematician and social scientist, the Marquis de Condorcet, in 1785.

Democratic voting seems to create a logical contradiction. As we pass from individual choices to some form of collective choice a paradox arises. Collective rationality does not seem to be merely the sum of individual rationalities.

Social choices are quite different beasts from individual choices, despite the fact that social choices are composed of individual choices. As a result, collective social choices sometimes exhibit an arbitrariness that does not reflect the way that personal decisions are made. Personal decisions arise from individual inclinations and tastes, but collective social choice does not. Society does not have an inclination or a taste of its own.

The most important modern discovery about the example of the Star Chamber is that it is not an artificially constructed scenario never likely to be encountered in real life.[52] The American political scientist Alan Taylor, has displayed an analogous type of problem which actually arose in a 1980 election for a place on the US Senate, representing New York.[53] The three candidates involved were a conservative, Alphonse D'Amato (who subsequently became well known as the head of the committee investigating the Whitewater affair involving the Clintons), and two liberals, Elizabeth Holtzman and Jacob Javits. All three parties conducted thorough exit polls which, we shall assume, give a reliable record of what the preferences of voters were for the three candidates. The six possible orders of preference for the three candidates were found with the percentages among voters from the exit polls, as shown in Table 8.1.

The official result of the election was close: D'Amato received 45 per cent of the vote, Holtzman 44 per cent and Javits just 11 per cent. But look at the results of a head-to-head contest according to the exit polls. Holtzman would have defeated Javits by 66 per cent to 34 per cent in a head-to-head contest, and would also have defeated D'Amato by 51 per cent to 49 per cent. Clearly the outcome of an election depends rather sensitively on how you handle the votes.

These paradoxes of rational choice display what logicians call *intransitivity*: the fact that A prefers B, and B prefers C does not mean that A prefers C. If A prefers B, and B prefers C *does* mean that A prefers C, then the situation is called

Table 8.1 US Senate election, New York 1980: exit polls

Place			Voting preference (percentage)
First	Second	Third	
D'Amato	Holtzman	Javits	22
D'Amato	Javits	Holtzman	23
Holtzman	D'Amato	Javits	15
Holtzman	Javits	D'Amato	29
Javits	Holtzman	D'Amato	7
Javits	D'Amato	Holtzman	4

transitive. We have seen that preference can be an intransitive relationship. Beating rival football teams is an intransitive relation: if Arsenal defeat Spurs and Spurs defeat Chelsea, then this does not mean that Arsenal will necessarily defeat Chelsea. 'Liking someone' is another intransitive relation: 'Peter likes Paul' and 'Paul likes Pippa' does not guarantee that 'Peter likes Pippa'. By contrast, a relation like 'being larger than' is transitive. If the number A is larger than B and B is larger than C then A must be larger than C. The paradoxical thing about making choices, we are going to discover, is that a concept of rationality based on transitivity cannot be transferred from individuals to collections of individuals by means of any reasonable rule for taking majority decisions.

Intransitivity can also emerge in other forms, for instance when electors vote for parties according to their stands on different issues. Suppose that voters do not have an opportunity to vote on single issues, but only for candidates who hold positions on two issues. The alternatives on the first issue might be 'state health care' (S) or 'private health care' (P), and on the second issue 'more jobs' (J) or 'lower taxes' (T). The four possible combined stances of the candidates on these two issues are SJ, ST, PJ, and PT. Suppose that the preferences of three voters on these four possible stances are ordered as

Voter 1: (SJ, ST, PJ, PT)
Voter 2: (ST, PT, SJ, PJ)
Voter 3: (PJ, PT, SJ, ST).

We see that there is no possible stance on the two issues that can defeat all the others if they are compared one on one. The preferences are therefore intransitive (see Figure 8.4).

What is tantalizing about this simple example is the fact that if we had separate votes on the two issues, then S would be preferred to P by voters 1 and 2, and J would be preferred to T by voters 1 and 3. Yet, despite the fact that a majority would favour S over P, and J over T, if there were separate votes on the two issues, the combined stance PT *defeats* SJ because it is supported by a majority of voters (voters 2 and 3). Remarkably, a majority stance on a combination of issues can be composed of alternatives that only minorities favour. This is why politicians like to construct agendas that appeal to groups of different minorities. Again, we see that the path from individual preferences to collective wishes can be unreliable and counter-intuitive.

These voting paradoxes have been known in simple forms since the nineteenth century, but for the most part were regarded as curiosities that could always be avoided in real life, rather like the logical paradoxes that preceded the developments in logic that we looked earlier in this chapter. Things took an analogous dramatic turn in 1950 when the American economist Kenneth Arrow

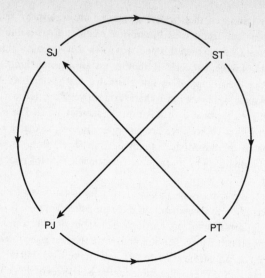

Fig. 8.4 There is no stance on the two issues that defeats all others. The arrows indicate the majority preferences between stances on the two issues. The preferences are cyclical and so the social preferences are intransitive.

analysed the problem of democratic choice in a general and transparent fashion. The result, which helped earn Arrow win the Nobel prize for economics in 1972, was published as an essay with the innocuous title 'A difficulty in the concept of social welfare'.[54] It is now known as the Arrow Impossibility Theorem by the pessimists, and as the Arrow Possibility Theorem by the optimists!

The Arrow Impossibility Theorem

In an autocracy, one person has his way; in an aristocracy a few people have their way; in a democracy no one has his way.

CELIA GREEN[55]

Arrow wanted to cut through the vast array of different possible voting systems by isolating the essential features of any democratic system in order to discover if there were any conditions under which intransitivity could be avoided. He assumed that the individual preferences satisfy two simple rules:

(a) *Comparability of alternatives.* If there are two alternatives, x and y, then either x is preferred to y or y is preferred to x. This requires the alternatives to have some property in common which can be used to compare their worth. Ties are not allowed in the individual preferences.

(b) *Transitivity*. Individual choices by voters are consistent in their order of preference; that is, if x is preferred to y and y is preferred to z then, necessarily, x is preferred to z. Note that we are trying to determine whether or not this property will be shared by the collective will of the voters.

The next step was to choose defining characteristics of democratic choice which one might wish any system of social choice derived from many individual choices to respect. Five were chosen:

Condition 1. *Unrestricted freedom of individual choice*
Each individual voter is allowed to choose any one of the possible orderings of the candidates. There are no organizations that can prevent some preferences being expressed by voters.

Condition 2. *Social choice should positively reflect individual choices*
If the social choice is that x is preferred to y and no individuals change their preference for x over y, then x must remain socially preferred to y. The changing of preferences for other alternatives should be irrelevant to the question of whether x is socially preferred to y. This ensures that the method of totalling up the individual votes to get the collective choice is not perverse.

Condition 3. *Irrelevant alternatives should have no effect*
The social ordering of some subset of choices is not altered by changes in the ordering of the other possibilities not in this subset.

Condition 4. *The voice of the people counts*
The outcome of the election is not imposed. The social choice cannot be unrelated to the individual voters' choices. This prevents the social choice being imposed upon society from outside, for example, by some religious belief.

Condition 5. *No dictatorship*
There is no individual such that this individual's choice always determines the overall social choice. This prevents the social choice being imposed upon society from within, by one individual.

The purpose of these conditions is to allow a rigorous examination of the consequences of many possible links between individual and collective choice, constrained only by fairly reasonable restrictions that most members of society would regard as desirable, if not essential, to democratic choice. Remarkably, Arrow proved that *if individual choices are finite in number, and obey the conditions (a) and (b), then there is no method of combining individual preferences to produce a social choice which meets all the conditions (1)–(5)*.

Every method of making a social choice that satisfies the conditions (1)–(3) either violates the conditions (4) or (5) or it contravenes the requirements (a) or (b). Note that the social intransitivity does not arise from any intransitivity of individual preferences, because they are explicitly forbidden by assumption (b). If the democratic conditions (1)–(5) and (a) are satisfied, then there must be intransitivity in the outcome. There is no such thing as a social consensus.

As with Gödel's Incompleteness Theorem, it is important to look at the assumptions that underlie Arrow's theorem to see where the weakest link in the chain lies. Arrow's conditions are all necessary in the sense that if any one of them is dropped the conclusion disappears with it. Thus, if one of the conditions is undermined, the conclusion that transitive social choice is impossible no longer holds. The most interesting condition to scrutinize more closely is (1): the unrestricted nature of individual choices.

Even before Arrow's work, it was known that a transitive majority decision could be obtained when individual preferences were restricted in some way. Work by Duncan Black,[56] which was developed further by Amartya Sen,[57] showed that a majority decision (one that is preferred by a majority of voters to all the others that are possible) is never possible when each alternative is ranked differently by each voter. This is evident in the simple 'Star Chamber' example given earlier, where every voter ranks each preference slot (first, second, third) differently. Situations in which each alternative is ranked differently by every voter always create intransitivity and paradox. When this is not the case, and individuals agree on the best, second best,..., and worst candidates, then the situation satisfies Arrow's conditions (a)–(b) and (2)–(4), although not condition (1). This reveals that Arrow's impossibility could be avoided if the preferences of voters exhibit some degree of similarity and there is a trend of public opinion. Sen called a set of voter preferences *value restricted* if all voters agree that there is some alternative that is never best, intermediate, or worst for every set of three alternatives (and analogously for any number of voters and alternatives).

Another reaction to these paradoxes is to hope and pray that they are very unlikely. If so, then the paradoxes could be dismissed as a harmless technicality. The likelihood is easy to determine using Sen's result that intransitivity arises when there are distinct preference orders for each voter. Suppose that three voters order three preferences: how likely it is that a paradoxical outcome arises? That is, one where A beat B, B beats C, but C beats A.

Each of the three voters has six ways of ordering their preferences (ABC, ACB, BAC, BCA, CAB, and CBA); so the three voters can express their preferences in a total of $6 \times 6 \times 6 = 216$ possible ways. A paradox arises only when the first, second, or third choices of any two voters do *not* agree. Consider how many ways there are for this situation to arise. After the first voter has made any one of the six available choices, the second voter has only two choices available that do have the same preference in first, second, or third place; that then leaves the third voter only one possible choice that differs in each place. So, the number of ways in which their choices can be made so as to create a logical paradox is $6 \times 2 \times 1 = 12$. Therefore, the probability of a paradox arising is just the fraction $12/216 = 0.056$ of the possible voter preferences that lead to paradox. This is 5.6 per cent.

If we increase the number of voters choosing between three alternatives, then the probability of paradox rises only slightly, gradually approaching 8.0 per cent. But, if the number of alternatives on offer to the voters increases, then the probability of paradox rises quickly to certainty (100 per cent). If the number of voters is an even number, then the possibility of ties arises, and there need not be a single winner preferred by all the voters. A tabulation of the probabilities of paradox arising for increasing odd numbers of voters and alternatives is shown in Table 8.2.[58]

Far from being a harmless technicality, the problem of paradox enshrined in Arrow's Impossibility Theorem seems to be ubiquitous and common. One must be a little careful in taking over these probabilities into real-life elections because they assume (for simplicity) that each choice is equally likely. In practice, this will not be the case, of course; many subjective factors will change voters' perception of the relative attractiveness of the different alternatives. However, in practice, sophisticated electorates (especially small ones within political institutions) may respond to the possibility of paradox by deliberately engineering it for their own party's advantage. The only voting systems that are strategy-proof are dictatorial.

Another way to sidestep the conclusions of Arrow's theorem is to replace the Dictator by a 'randomizer'—some arbitrary way of imposing a social choice when the situation is intransitive. We often do this when we are making choices,

Table 8.2 Probabilities of no majority winner

Number of alternatives	Number of voters						
	3	5	7	9	11	...	Limit
3	0.056	0.069	0.075	0.078	0.080	...	0.080
4	0.111	0.139	0.150	0.156	0.160	...	0.176
5	0.160	0.200	0.215	0.230	0.251	...	0.251
6	0.202	0.255	0.258	0.285	0.294	...	0.315
7	0.239	0.299	0.305	0.342	0.343	...	0.369
.	.	.	.	.	.	.	
Limit	1.000	1.000	1.000	1.000	1.000	...	1.000

The probability of a voting paradox arising from intransitivity as the number of voters and voter choices changes. As the number of voters choosing between three alternatives increases, the probability of paradox approaches 8 per cent. As the number of alternatives increases, the probability of paradox approaches 100 per cent for any number of voters.
Data from S. Brams, *Paradoxes in Politics* (1976).

whether it is by just taking the plunge and picking one of the options, or by explicitly introducing a randomizer (tossing a coin, drawing lots, etc) to overcome the lack of a rationally clear choice or to break a deadlock. Frank Tipler has suggested that it may be necessary to incorporate a randomizer as a sublevel within any 'mind' (human or artificial) in order to break occasional deadlocks in choice created by intransitivities.[59] This randomizer might be linked to the underlying quantum uncertainties which some, like Roger Penrose, have claimed can play a role in neural information processing.[60] If breaking these deadlocks is important for decisive action, then this would be a feature that would be an aid to survival, and should therefore be adaptive.

These results are striking. They show how counter-intuitive can be the behaviour of complex systems. They also have something to say about reductionistic explanations of systems like the human mind. We have seen how collective choices need not be simply related to individual choices (whether they are made by human voters placing crosses on pieces of paper, computer switching, or quantum randomness). Curiously, the trend towards a future in which individual choices might be summed almost instantaneously to give members of democracies more choice in the way they are governed, or in the products available to them, makes the future less rational in some deep way unless particular restrictions are placed upon the choosers or their range of choices.

Summary

Secant, cosine, tangent, sine
Logarithm, logarithm
Hyperbolic sine
3 point 14159
Slipstick, sliderule
TECH TECH TECH!

CAL TECH 'BEAVERS' CHEER

In this chapter, we have focused upon some of the ways in which various types of impossibility and unpredictability affect us. In some cases they limit our ability to predict or offer a future horizon of impossibility which will bring the curtain down on our efforts to understand the Universe. But the situation with regard to Gödel's theorem is far from as simple as earlier commentators have implied. We saw how the fine print of Gödel's theorem allows all manner of different conclusions to be drawn about its impact (or non-impact) upon the scientific enterprise. Many scientists have sought to use Gödel's ideas to place restrictions upon the scope of computers to do what the human mind does. So far, these do not seem to be entirely persuasive. Computers will be able to frame the same arguments about human minds. Only by taking seriously the fallibility of human minds are we able to distinguish their scope from that of artificial intelligences. The contributions of Turing, Popper, and Mackay take us further into the psychological consequences of these results, showing how they shed new light upon the cogency of the famous problem of free will. Along the way we discovered that a much-quoted theorem of mathematical politics regarding the predictability of elections is in fact false. Finally, we have probed further into the realm of the social sciences to explore the strange impossibility that Arrow has discovered in rational voting systems. The process of passing democratically from individual to collective choices is doomed by impossibility: there is no reliable way of establishing rational collective choices. While these paradoxes were established by considering voting systems within political and economical contexts, they have fascinating applications to some theories of the workings of the human mind, which see it acting as a 'society' of neuronal voters.

CHAPTER 9

Impossibility: taking stock

Everything is deemed possible except that which is impossible in the nature of things.
CALIFORNIA CIVIL CODE

Telling what is from what isn't

man is not a circle with a single centre; he is an ellipse with two foci.
Facts are one, ideas are the other.
VICTOR HUGO, *LES MISERABLES*

The idea that some things may be unachievable or unimaginable tends to produce an explosion of knee-jerk reactions amongst scientific (and not so scientific) commentators. Some see it as an affront to the spirit of human inquiry: raising the white flag to the forces of ignorance. Others fear that talk of the impossible plays into the hands of the anti-scientists, airing doubts that should be left unsaid lest they undermine the public perception of science as a never-ending success story. Finally, there are those who seize upon any talk of impossibility as an endorsement of their scepticism about unbridled technological progress, tramping roughshod over the environment and human dignity: the unstoppable in pursuit of the unsustainable.

If this book has taught the reader anything, I hope it is that the notion of impossibility is far subtler than naïve assumptions about the endless horizons of science, or pious hopes that boffins will be baffled, would lead you to believe. Limits are ubiquitous. Science exists only because there are limits to what Nature permits. The laws of Nature and the unchanging 'constants' of Nature define the borders that distinguish our Universe from a host of other conceivable worlds where all things are possible. In those imaginary worlds of unlimited possibility there can exist neither complexity nor life. They contain no imaginations. The fact that we can conceive of logical and practical impossibilities is a reflection of a self-reflective consciousness that is so far unique among the fellow creatures with whom we share our planet. Impossibilities allow those conscious complexities to exist.

For thousands of years the exploration and creation of impossibility in language and art has provided the mind with a stimulating virtual environment in which to exercise its penchant for association and rationalization. Philosophers have wrestled with concepts that straddle the border between possibility and impossibility. Theologians have struggled to reconcile the concept of a Being for whom nothing is impossible with the necessities of logic and the laws of Nature.

We have seen how many of these contemplations have seeded profound developments in the way we think about the Universe. The artistic creation of impossible figures has produced new insights into the workings of the mind. Like all art, this creates safe alternative realities for our minds to explore. From the tantalizing circularities of linguistic and logical paradox have emerged deep discoveries about the nature of logic and mathematics. We have learned that those logical structures which are complex enough to express truths about themselves cannot be fully captured by predictable lists of rules and axioms.

In every area of human competence we have made significant progress. This progress has been most evident in the profusion of technological gadgets that surround us. Here, human ingenuity has found a treasure trove of possibilities. It is hard to imagine that the well will ever run dry. It is easy to presume that this productive progress will be never-ending and all-defining for the human enterprise; easy to see the nature of reality as the sum total of what is technically possible; easy to regard the boundaries as ever-receding irrelevancies. We explored some of the ways in which the scientific endgame might be played out; saw some of the extraordinary coincidences that must exist if our capability is to remain a match for the subtlety of Nature; and located ourselves on the ladder of progress that defines our present capability to manipulate things larger, smaller, and more complex than ourselves.

Limits to what is impossible may turn out to define the Universe more powerfully than the list of possibilities. On a variety of fronts we have found that growing complexity ultimately leads to a situation that is not only limited, but self-limiting. Time and again, the development of our most powerful theories has followed this path: they are so successful that it is believed that they can explain everything. Commentators begin to look forward to the solution of all the problems that the theory can encompass. The concept of a 'theory of everything' occasionally rears its head. But then something unexpected happens. The theory predicts that it cannot predict: it tells us that there are things that it cannot tell us. Curiously, it is only our most powerful scientific theories that seem to possess this self-critical feature.

We have explored a range of limits that exist on our quest to understand the Universe of which we are a part. There are human limits, which arise from the nature of our humanity and the evolutionary inheritance we all share. There are technological limits, created at root by our biological nature. Our limited size

and strength, together with the temperate nature of our bio-friendly environment, force us to pursue technological progress: to devise artificial means of
probing the extremes of size, complexity, and temperature that are possible in
the world around us. In all these quests we encounter unexpected limits on what
can be done. Information is expensive to acquire. It costs time and energy. There
are limits to the speed at which that information can be transmitted, and limits
to the accuracy with which it can be specified or retrieved. But most important
of all, there are powerful limits on how much information can be processed in
reasonable periods of time. We are surrounded by a host of practical problems,
too complicated for the human brain to solve unaided, which even the fastest
computers that Nature allows cannot solve. These problems are intractable.
Many of them sound simple, but their solution requires more space and time
than the entire Universe allows.

These limits are boundaries brought about by practicalities, costs, and time.
Some are just extrapolations of our everyday experience. All might be steadily
pushed back in the far future. But we found that there exist unexpected limits
which define more fundamental levels of impossibility. The further we go from
the everyday realm of human experience in our quest to understand the nature
of the Universe, the more surprising are the limits we encounter.

The astronomers' desire to understand the structure of the Universe is
doomed merely to scratch the surface of the cosmological problem. All the
great questions about the nature of the Universe—from its beginning to its end—
turn out to be unanswerable. There is a fundamental divide between the part of
the Universe that we can observe and the entire, possibly infinite, whole. There is
a visual horizon beyond which we cannot see or know. Again, there is a positive
side to this limitation. If it did not exist, then nor would we: every movement
of every star and galaxy would be instantly felt, here and now.

Until quite recently scientists believed it reasonable to assume a 'what you see
is what you get' theory of the Universe: that what there is of the Universe beyond
our horizon is the same, on average, as the part that we can see. Unfortunately,
our most compelling theories of the evolution and structure of the Universe
sweep away these simple expectations: we expect the Universe to be endlessly
diverse both geographically and historically. It is most unlikely to be everywhere, even roughly, the same. We are more likely to inhabit a little island of
temperate tranquillity amid a vast sea of cosmic complexity, for ever beyond our
power to observe.

The speed at which light travels is limited and so, therefore, is our knowledge
of the structure of the Universe. We cannot know whether it is finite or infinite,
whether it had a beginning or will have an end, whether the structure of physics
is the same everywhere, or whether the Universe is ultimately a tidy or an untidy
place.

As we moved on to explore impossibilities that lie more deeply embedded in the nature of things, we found that logic and mathematics are incestuously stricken by limits on their power to predict and explain. Just as a simple game, like noughts and crosses, is completely predictable for perfect players, so very simple logical structures can be fully understood. But as a logical structure is made more complex, there is a sudden change. When it reaches a particular critical level of complexity it becomes impossible to understand it fully: impossible to show that it is self-consistent. That critical level is strikingly low: it is identified by the presence of the familiar arithmetic of numbers that is embedded deep within human intuition, and which has sufficed to understand the complexities of the physical world around us.

These deep limitations spread out into the realms of computation, mathematical deduction, and the assessment of complexity and randomness. They have appeared to be so all-embracing that many people have sought to find consequences for workings of the human mind or the quest to fabricate an artificial intelligence surpassing our own. Others see these deep limits as the ultimate insurance policy against any full understanding of Nature's laws. For if mathematics cannot capture all truth within a finite set of rules, it will surely not be possible for physicists to capture the workings of physical reality in a finite collection of laws of Nature? This argument is a leap too far. We saw how the small print of Gödel's famous incompleteness theorems is important. Like the microscopic conditions in feint ink on the reverse of your insurance policy, they are the most important ingredients. If they do not apply, then neither does the headlined guarantee.

In the last chapter, we traced the possible implications of these deep forms of impossibility for some aspects of the human mind. We looked at free will and determinism, and learned why computers and minds cannot fully understand themselves, or predict their own futures. Time travel challenges us to conceive of worlds which are not only unpredictable but inconsistent. We saw how many common paradoxes of time travel hide confusions rather than contradictions. Finally, we encountered puzzling impossibility in any voting process. Whether it be an election, a bank of linked computers, or the 'voting' neurones inside our brains, it is impossible to translate individual rational choices into collective rationality. Again, we see a threat to our confident extrapolations about the behaviour of complex collective intelligences in the far distant future. Our experience of complex systems is that they display a tendency to organize themselves into critical states that are optimally sensitive, so that small adjustments can produce compensating effects throughout the system. As a result, they are unpredictable in detail. Whether it is sand grains or thoughts that are being self-organized, their next move is always a surprise.

We live in strange times. We also live in strange places. As we probe deeper into the intertwined logical structures that underwrite the nature of reality, I believe that we can expect to find more of these deep results which limit what can be known. Our knowledge about the Universe has an edge. Ultimately, we may even find that the fractal edge of our knowledge of the Universe defines its character more precisely than its contents; that what cannot be known is more revealing than what can.

Notes

I love being a writer. What I can't stand is the paperwork..

PETER DE VRIES

Chapter 1

[1] W.H. Auden, 'Reading', *Dyer's Hand* (1963).

[2] J.M. Barrie, *The Admirable Crichton*, act 1 (perf. 1902, publ. 1914).

[3] For a scholarly account of the Pythagoreans, see J.A. Philip, *Pythagoras*, University of Toronto Press, Toronto (1966); for a more popular account, see P. Gorman, *Pythagoras: A Life*, Routledge, London (1979).

[4] See the three classic collections of short stories by J.L. Borges, *Labyrinths*, New Directions Press, New York, 2nd edn (1964), *The Aleph and other Stories 1933–1969*, Dutton, New York (1978), and *The Book of Sand*, Penguin Books, London (1979).

[5] J.D. Barrow, *The Artful Universe*, p. 62, Oxford University Press (1995).

[6] This is an 'average' composite female face created in 1997 by K. J. Lee, D. A. Rowland, D. I. Perrett and D. M. Burt of the School of Psychology, University of St. Andrews.

[7] J.D. Barrow and S.P. Bhavsar, What the astronomer's eye tells the astronomer's brain, *Quarterly Journal of the Royal Astronomical Society*, **28**, 109 (1987).

[8] E. De Bono, *A Five-day Course in Thinking*, Penguin Books, London (1968).

[9] For a detailed discussion of the Game of Life, see E. Berlekamp, J.H. Conway, and R. Guy, *Winning Ways*, Academic Press, New York (1982); a lower-level discussion can be found in W. Poundstone, *The Recursive Universe*, Morrow, New York (1985).

[10] Matthew 19, v. 21.

[11] Medieval theologians grappled seriously with such dilemmas. For example, by the middle of the thirteenth century a number of fine distinctions had been made between things innate to divine power and things that the Deity had chosen to do. There were things that God has the capacity to do (although had not done), things that he does not do, and things that he cannot do. Gradually, the theological emphasis shifted subtly to talk about different ways of speaking of divine powers rather than of the different powers themselves. The doctrine was important to theologians because they needed to affirm God's freedom of action without undermining the reliability or the God-ordained appropriateness of the world around them. Further discussion can be found in the collection of articles by T. Rudavsky (ed.), *Divine Omniscience and Omnipotence in Medieval Philosophy*, Reidel, Dordrecht (1985); see also A. Kenny, *The God of the Philosophers*, Oxford University Press (1978); J.F. Ross, *Philosophical Theology*, Bobbs-Merrill, Indianapolis (1979); N. Kretzmann, Omniscience and immutability, *Journal of Philosophy*, **63**, 409 (1966); J. Wippel, The reality of non-existing possibles, *Review of Metaphysics*, **34**, 729 (1981); and K. Ward, *Religion and Creation*, Clarendon Press, Oxford (1996).

[12] H. Pagels, *The Dreams of Reason*, p. 286, Simon and Schuster, New York (1988).

[13] J. Polkinghorne, *One World: the interaction of science and theology* SPCK, London (1994); A. Peacocke, *God and the New Biology*; and J. Doye, I. Goldby, C. Line, S. Lloyd, P. Shellard, and D. Tricker, Contemporary Perspectives on Chance, Providence and Free Will, *Science and Christian Belief*, 7, 117 (1995).

[14] T. Brown, *Religio Medici*, Vol. 1, p. 47, Dutton, New York (1934), (London 1658).

[15] Various counting taboos are described in J.D. Barrow, *Pi in the Sky*, Oxford University Press (1992); see also C. Panati, *Sacred Origins of Profound Things*, Penguin Books, London (1996) and W. Buckert, *Creation of the Sacred*, Harvard University Press, (1996).

[16] Genesis 2, v. 9.

[17] The encryption system Pretty Good Privacy, or PGP, written by Philip Zimmermann is available on a website http://www.arc.unm.edu/~drosoff/pgp/pgp.htm#whatispgp.

[18] A Cromer, *Uncommon Sense*, p. 78, Oxford University Press (1993).

[19] For further discussion on the connections between monotheism and the concept of laws of Nature, see J.D. Barrow, *The World Within the World*, Oxford University Press (1988); J. Needham, *The Grand Titration, Science and Society in East and West*, Allen and Unwin, London (1969); J. Needham, *Human Law and the Laws of Nature in China and the West*, Oxford University Press (1951); and F. Oakley, Christian theology and the Newtonian science: rise of the concept of laws of Nature, *Church History* 30, 433 (1961).

[20] J. Needham, *Science and Civilisation in China*, vols 1–7, Cambridge University Press (1954–).

[21] O. Wilde, *Phrase and Philosophies for the Use of the Young*, (1891).

[22] S. Brams, *Superior Beings: if they exist, how would we know?* Springer, New York (1983).

[23] N. Falletta, *The Paradoxicon*, p. xvii, Doubleday, New York (1983).

[24] See P. Hughes and G. Brecht, *Vicious Circles and Infinity: an anthology of paradoxes*, Penguin Books, London (1978).

[25] A. Rapoport, Escape from Paradox *Scientific American*, July 1967, pp. 50–6.

[26] The Independent newspaper, London, p. 15, 19 July 1997.

[27] B. Ernst, *The Magic Mirror of M.C. Escher*, Tarquin, Norfolk (1985).

[28] B. Ernst (a.k.a. J.A.F. Rijk), *The Eye Beguiled: Optical Illusions*, p. 69, Taschen, Cologne (1992).

[29] L.S. Penrose and R. Penrose, Impossible objects a special type of visual illusion, *British Journal of Psychology*, 49, 31 (1958).

[30] W. Hogarth, Frontispiece to John Joshua Kirby, *Dr. Brook Taylor's Method of Perspective Made Easy*, London (1754).

[31] Original print in the Philadelphia Museum of Art.

[32] These can be viewed on the website http://www.biggallery.com/art/byartist/Z1004689.asp.

[33] B. Ernst, ref. 5, p. 68.

[34] L. Necker, Observations on some remarkable phenomena seen in Switzerland: and on an optical phenomenon which occurs when viewing a figure of a crystal or geometrical solid, *London and Edinburgh Philosophical Magazine and Journal of Science*, 1, 329–37 (1832); H. Barlow, The coding of sensory messages, in W. Thorpe and O. Zangwill (eds), *Current Problems in Animal Behaviour*, pp. 331–60, Cambridge University Press (1961); R. Gregory, *The Intelligent Eye*, McGraw-Hill, New York (1970).

[35] Original in the Israel Museum Collection, Jerusalem.

[36] See ref. 5.

[37] H.S.M. Coxeter, Four-dimensional geometry in *Introduction to Geometry*, pp. 396–412, Wiley, New York (1961); B. Grünbaum and G.C. Shepherd, *Tilings and Patterns*, W.H. Freeman, New York (1987).

[38] See ref. 5 and J.L. Borges, S. Ocampo, and A.B. Casares (eds) *The Book of Fantasy*, Black Swan, London (1990).

[39] G. Vlastos, Zeno of Elea, in *Encyclopedia of Philosophy*, Vol. 8, pp. 369–79, Macmillan, New York (1967).

[40] For modern discussions of Zeno, see A. Grünbaum, *Modern Science and Zeno's Paradoxes*, Wesleyan, Middletown (1967) and W.C. Salmon (ed.), *Zeno's Paradoxes*, Bobbs-Merrill, Indianapolis (1970).

[41] E. Taylor and J.A. Wheeler, *Spacetime Physics*, W.H. Freeman, New York (1966).

[42] A. Rae, *Quantum Physics—Illusion or Reality*, Cambridge University Press (1986). J. Gribbin, *In Search of Schrödinger's Cat*, Bantam, New York (1984).

[43] B. d'Espagnet, *In Search of Reality*, Springer, New York (1983); D. Mermin, Is the Moon there when nobody looks?, *Physics Today*, p. 38 (April 1985); P.C.W. Davies, and J.R. Brown, (eds), *The Ghost in the Atom*, Cambridge University Press (1986).

[44] H.R. Brown and R. Harré, *Philosophical Foundations of Quantum Field Theory*, Oxford University Press (1988).

[45] E. Wigner, Remarks on the mind–body question, in I.J. Good (ed.), *The Scientist Speculates: an anthology of partly-baked ideas*, p. 284, Basic Books, New York (1962).

[46] Titus 1, v. 12.

[47] P.V. Spade, The Medieval Liar: a Catalogue of the Insolubilia Literature, Pontifical Institute, Toronto (1975).

[48] B. Russell, *The Principles of Mathematics*, 2nd edn, Norton, New York (1943).

[49] D. Adams, *The Restaurant at the End of the Universe*, Ballentine, New York (1995).

[50] B. D'Espagnet, The quantum theory and reality, *Scientific American*, Nov. 1979, p. 158; N. Herbert, *Quantum Reality*, Rider, London (1985); D. Lindley, *Where the Weirdness Goes*, Basic Books, New York (1996).

[51] K. Wilber (ed.), *Quantum Questions: Mystical Writings of the World's Great Physicists*, Shambhala, Boston (1985).

[52] G. Edelman, *Bright Air, Brilliant Fire; On the Matter of the Mind*, Penguin Books, London (1992).

[53] A paperback reprint is available as G. Gamow, *Mr Tompkins in Paperback*, Cambridge University Press (1965).

[54] E. Wigner, The unreasonable effectiveness of mathematics in the natural sciences, *Communications on Pure Applied Mathematics*, 13, 1 (1960).

[55] Particle Data, *Reviews of Modern Physics* 54. 1 (1996).

[56] See ref. 40.

[57] CERN Courier, July/August (1997), p.22.

Chapter 2

[1] P.W. Frey (ed.), *Chess Skill in Man and Machine*, 2nd edn, Springer, New York (1983). M. Newborn, *Kasparov versus Deep Blue: Computer Chess Comes of Age*, Springer, New York, (1997); The use of chess as an image of Nature appears first in T. H. Huxley's *A*

Liberal Education (1868). He writes that, 'The chess board is the world, the pieces are the phenomena of the universe, the rules of the game are what we call the Laws of Nature. The player on the other side is hidden from us. We know that his play is always fair, just and patient. But also we know, to our cost, that he never overlooks a mistake, or makes the smallest allowance for ignorance'.

2   Limits are possible, for example, on the existence of superweak forces. The biggest open question is still whether the gravitational force has another part that increases linearly with distance. This component is associated with the possibility of a new constant of Nature, the 'cosmological constant', but astronomical evidence is unable to tell us as yet whether it has a non-zero value.

3   See the example M. Gell-Mann, *The Quark and the Jaguar*, Little Brown, New York (1994).

4   Here, one needs a word like mañana but which, as my colleague Leon Mestel once remarked, does not convey the same degree of urgency.

5   J. Ortega y Gasset, *The Revolt of The Masses*, Mentor Books, New York (1950).

6   Addition by Orson Welles to Graham Greene's text of *The Third Man* for the 1949 screenplay.

7   G. Stent, *The Coming of the Golden Age: A View of the end of Progress*, Natural History Press, New York (1969). A later book by Stent, *Paradoxes of Progress*, W.H. Freeman, San Francisco (1978), reprints the first three chapters of his earlier work together with further essays on the future of biology and science in general. There were other authors who identified the trends of the 1960s and the beatnik generation with the degeneration of human culture over a very wide spectrum; see, for example, O. Guiness, *The Dust of Death*, Inter-Varsity Press, London (1973).

8   J. Horgan, *The End of Science*, Addison-Wesley, Reading, Mass. (1996).

9   G. Stent, *The Coming of the Golden Age*, p. xi.

10 This view receives an interesting modern endorsement by the novelist and travel writer Paul Theroux in his account of his travels in the Pacific Islands, told in *The Happy Isles of Oceania: Paddling the Pacific*, Penguin Books, New York (1992).

11 G. Stent, *The Coming of the Golden Age*, p. 132.

12 See J.D. Barrow and F.J. Tipler, *The Anthropic Cosmological Principle*, Clarendon Press, Oxford (1986) for some discussion of this group.

13 Of course, some social scientists have already attempted to apply this interpretation to the whole of contemporary science, not merely to the speculative frontiers that Horgan highlights. However, most scientists regard this surprisingly prevalent view as absurd. A vigorous critique of it has been provided recently by Steven Weinberg. This article was stimulated by an amusing Trojan horse attack by the physicist Alan Sokal on the level of critical intelligence being exhibited in some areas of the humanities. An absurd spoof article by Sokal entitled 'Transgressing the Boundaries: Toward a Transformative Hermeneutics of Quantum Gravity' was approved for publication by the editors of *Social Text*, a leading journal. When this came to light it sparked a fierce debate over a spectrum of issues, ranging from ethics to the competence of some social commentators to say anything about science at all. The article appears in vol **14**, 62–4 (1996).

14 J. Horgan, op. cit., p. 7.

15 P.C.W. Davies, *The Mind of God*, Simon and Schuster, New York (1992).

16 H. Weyl, *God and the Universe: The Open World*, p. 28, Yale University Press, New Haven (1932).

[17] An interesting remark about the relation between religion and science fiction is made by James Gunn, in *The New Encyclopedia of Science Fiction*, Viking, New York (1988): 'Science fiction, like science, is an organised system that, for many, takes the place of religion in the modern world by attempting to complete explanation of the universe. It asks the questions—where did we come from? why are we here?—where do we go from here?—that religions exist to answer. That is why *religious* science fiction is a contradiction in terms although science fiction *about* religion is commonplace.' While a slightly inaccurate (recall C.S. Lewis's science fiction trilogy with a religious aspect), this is an interesting perspective.

[18] For a classic study from the first half of this century, see J.B. Bury, *The Idea of Progress*. Macmillan, New York (1932) and reprint by Dover, New York (1955). See also R. Nisbet, *History of the Idea of Progress*, Heinemann, London (1980) and E. Zilsel, The genesis of the concept of scientific progress, *Journal of the History of Ideas*, 6, 325 (1945).

[19] There are many studies of the history of the cyclic universe; see, for example, M. Eliade, *The Myth of the Eternal Return*, Pantheon, Kingsport (1954). The Gifford Lectures of S. Jaki, *Science and Creation*, Scottish Academic Press, Edinburgh (1974), while providing many facts, are so biased and vitriolic in their interpretation that they cannot be recommended as a route into these questions.

[20] For a detailed account of these Design Arguments, see J.D. Barrow and F.J. Tipler, *The Anthropic Cosmological Principle*, Clarendon Press, Oxford (1986 and 1996).

[21] The human eye was a classic example cited by proponents of the naïve forms of the Design Argument;' see ref. 20. The American biologist George Williams has discussed the many ways in which the design of the human eye (like many other natural adaptations that have evolved by the step-by-step process of natural selection) is flawed and could be improved upon; see G. Williams, *Plan and Purpose in Nature*, Orion, London (1996).

[22] H. Spencer, *Principles of Ethics*, Williams and Norgate, London (1892–3).

[23] For example John Desaguliers; *The Newtonian System of the World, the Best Model of Government* (1728). See also J.D. Barrow, *The World Within the World*, p. 74, Oxford University Press (1988).

[24] F. Manuel, *The Religion of Isaac Newton* Clarendon Press, Oxford (1974) and *A Portrait of Isaac Newton*, Frederick Muller, London (1980).

[25] G. Sarton, *The Study of the History of Science*, Dover New York 1957.

[26] K. Lorenz, *Behind the Mirror*, Harcourt, Brace, Jovanovich, New York (1977) and Kant's Doctrine of the a priori in the light of contemporary biology, in *Yearbook of the Society for General Systems Research*, vol. VII, pp. 23–35, Society for General Systems Research, New York (1962); also reprinted in R.I. Evans, *Konrad Lorenz: the Man and His Ideas*, pp. 181–217, Harcourt, Brace, Jovanovich, New York. Both are translations of the German original, first published in 1941.

[27] B. de Spinoza, *Ethics*, (1670) in *Britannica Great Books*, Vol. 31, W. Benton, Chicago (1980).

[28] J. Richards, The reception of a mathematical theory: non-Euclidean geometry in England 1868–1883, in *Natural Order: Historical Studies of Scientific Culture*, B. Barnes and S. Shapin (eds), Sage Publications, Beverly Hills (1979); E.A. Purcell, *The Crisis of Democratic Theory*, University of Kentucky Press, Lexington (1973).

[29] Bob Dylan, Desolation Row, from *Highway 61 Revisited*, CBS SBPG 62572.

[30] A. Crombie, Some attitudes to scientific progress, ancient, medieval, and modern, *History of Science*, 13, 213 (1975).

[31] Memorably defined by Ambrose Bierce in *The Devil's Dictionarty* (Dover, New York (1958), originally published 1911), as 'a philosophy that denies our knowledge of the Real and affirms our ignorance of the Apparent. Its longest exponent is Comte, its broadest is Mill, and its thickest is Spencer.'

[32] G. Lenzer, *Auguste Comte and Positivism: the essential writings*, Harper, New York (1975); L. Laudan, Towards a reassessment of Comte's *Methode Positive*, *Philosophy of Science*, **38**, 35 (1971).

[33] A. Comte, *Introduction and Importance of Positive Philosophy*, ed. F. Ferré, Bobbs-Merrill Co., Indianapolis (1976), p. 2.

[34] ibid., p. 2.

[35] ibid. p. 3.

[36] ibid. p. 3.

[37] ibid. pp. 31–2.

[38] Brush, p. 8; A. Comte, *System of Positive Polity*, vol. 1, pp. 312–3, transl. J. Bridget, Longmans, London (1851).

[39] J. Hervival, Aspects of French theoretical physics in the nineteenth century, *British Journal for the History of Science*, **3**, 109 (1966).

[40] S. de Laplace, *Philosophical Essay on Probabilities*, (1814), transl. F. Truscott and F. Emory, Dover, New York, (1951).

[41] The first lecture appeared in English translation as E. du Bois-Reymond, The limits of scientific knowledge, *Popular Scientific Monthly*, **5**, 17 (1874). The two lectures were published together as *Über Die Grenzen ds Naturerkennens: Die Sieben Welträtsel—Zwei Vorträge*, Leipzig (1916).

[42] In the years immediately before there had been considerable progress in understanding the laws of thermodynamics, motivated in part by a desire to understand the efficiency of heat engines, an integral part of the industrial revolution. These ideas had been applied, in theory, to the evolution of the Universe as a whole to deduce (erroneously it now turns out) that the Universe will gradually approach a global equilibrium of uniform temperature, termed 'the Heat Death' of the Universe. This inspired a good deal of philosophical pessimism; for further cosmological implications see J.D. Barrow, *The Origin of the Universe*, Orion, London (1994).

[43] E. Haeckel, *The Riddle of the Universe—at the close of the nineteenth century*, trans. J. McCabe, London and New York (1901) and reprinted as *The Riddle of the Universe*, Watts and Co., London (1929) as the third volume in the prestigious 'Thinker's Library' series.

[44] Haeckel, op. cit., pp. 365–6.

[45] See N. Rescher, *Peirce's Philosophy of Science*, University of Notre Dame Press, London (1978) for an excellent study of Peirce's ideas.

[46] Peirce expressed this as the mathematical limit of a convergent series of successive approximations to the truth.

[47] M. Planck, *Vortäge und Erinnerungen*, 5th edn, p. 169, Stuttgart, (1949), quoted by N. Rescher in *Scientific Progress: a philosophical essay on the economics of research in natural science*, Blackwell, Oxford (1978), p. 24.

[48] A.A. Michelson, cited in *Physics Today*, **21**, 9 (1968) and *Light Waves and their Uses*, University of Chicago Press, Chicago, (1961).

[49] G.B. Shaw, Maxims for Revolutionists: Reason, *Man and Superman*, Dodd, Mead & Co, New York (1939), (Westminster 1903).

Chapter 3

[1] W.E. Gladstone, House of Commons speech on the Reform Bill, 1866.

[2] Quoted in D. Michie (ed.) *Machine Intelligence*, vol 5, p.3 (1970).

[3] For an optimistic view of the scope of science, see the essay 'The limits of science' in P. Medawar, *The Limits of Science*, Oxford University Press (1984). However, Medawar's discussion is very limited and is especially weak with regard to the non-biological sciences. One of his unsurprising conclusions is that 'there is no limit upon the power of science to answer questions of the kind science can answer'!

[4] R. Penrose, *The Emperor's New Mind*, Oxford University Press (1989) and *Shadows of the Mind*, Oxford University Press (1994).

[5] Unfortunately, this claim rests upon the hidden assumption that the brain is infallible, and there is no reason to believe that it is. If we accept the brain's fallibility as an inevitable consequence of its evolution by a long and tedious process of natural selection, then we can no longer apply Gödel's theorem. Alan Turing's response to the situation was to claim that 'if a machine is expected to be infallible, it cannot also be intelligent. There are several theorems which say almost exactly that. But these theorems say nothing about how much intelligence may be displayed if a machine makes no pretence of infallibility', quoted by R. Penrose in *The Large, the Small and the Human Mind*, Cambridge University Press (1997), p. 112.

[6] M. Pepper (ed.) *The Pan Dictionary of Religious Quotations*, Pan, London (1991), p. 251.

[7] P. Duhem, *The Aim and Structure of Physical Theory*, Princeton University Press (1954), pp. 38–9. Duhem was an instrumentalist with a different type of non-realist view. For him, theories were tools for carrying certain tasks so he defined scientific progress by the success of theories in this respect.

[8] B. Glass, Science: endless horizons or golden age, *Science*, 171, 23–9 (1971) and Milestones and rates of growth in the development of biology, *Quarterly Review of Biology*, 54(1), 31 (1979).

[9] V. Bush, *Endless Horizons*, Public Affairs Paper, Washington (1990) reprint, Ch. 17: The Builder.

[10] K. Popper, *Objective Knowledge* Oxford University Press (1972), pp. 262–3.

[11] Quoted in I. Stewart, *The Problems of Mathematics*, Oxford University Press (1987).

[12] M. Foster, The growth of science in the nineteenth century, *Annual Report of the Smithsonian Institution* for 1899, Washington (1901), cited in Rescher, *Scientific Progress*, Blackwell, Oxford (1978), p. 49.

[13] D. Stauffer, *Introduction to Percolation Theory*, Taylor and Francis, London (1985).

[14] E. Witten, quoted in K. Cole, A Theory of Everything, New York Times Magazine, 18th October, 1987, p. 20.

[15] T. Kuhn, *The Structure of Scientific Revolutions*, 2nd enlarged edn., Univ. Chicago Press, (1970).

[16] S.W. Hawking, Lucasian lecture, delivered on 29 April 1980, reprinted in *Physics Bulletin*, Jan. 1981, pp. 15–17.

[17] This picture is based on an unpublished idea suggested by David Ruelle, the French mathematician who led the detailed study of what is known as 'chaos' in the 1970s, and was co-inventor of the term 'strange attractor'; see *Complexity*, 3 (1), 26 (1997).

[18] C. Sagan, *Contact: A Novel*, Arrow, London, (1985).

[19] I. Kant, *Prolegomena to Any Future Metaphysics*, (1857), quoted in N. Rescher, *Scientific Progress*, p. 248.

[20] A specific example is provided by the situation in elementary particle physics, where the standard theory of what is expected to occur at very high energies predicts a huge range of energies, called 'the desert', in which nothing new will be found.

[21] G. Priest, *Beyond the Limits of Thought* p. 6, Cambridge University Press (1995).

[22] G. Edelman. *Bright Air, Brilliant Fire*, Penguin Books, London, (1992) and *Neural Darwinism: The Theory of Neuronal Group Selection*, Basic Books, New York (1987). Not all workers in the field embrace this model. Francis Crick describes it, less impressively, as 'neural Edelmanism'.

[23] R. Penrose, *The Emperor's New Mind*, op. cit. and D.V. Nanopoulos, Theory of brain function, quantum mechanics and superstrings, CERN preprint CERN–TH/95–128 (1995).

[24] K. Devlin, *Goodbye Descartes*, Wiley, New York (1997); D. Dennett, *Kinds of Minds*, Orion, London (1996).

[25] W. Kneale, Scientific revolutions forever? *British Journal for the Philosophy of Science*, **19**, 27 (1967).

[26] J. Leslie, *End*, Routledge, London (1996).

[27] J.D. Barrow and F.J. Tipler, *The Anthropic Cosmological Principle*, ch. 10, Clarendon Press, Oxford (1986).

[28] To see this, note that the sum of the first four terms is bigger than four quarters ($= 1$), the sum of the next eight terms is bigger than eight eighths, and so on for ever.

[29] See D. Bohm, *Causality and Chance in Modern Physics*, Routledge, London (1957) for discussion of 'qualitative infinity of nature'.

[30] E. Wigner, The limits of science, *Proceedings of the American Philosophical Society*, **94**, 424 (1950).

[31] C. Babbage, *On the Economy of Machinery and Manufactures*, pp. 386–90, London (1835).

[32] D. Diderot, *Oeuvres complètes*, ed. J. Assezat, Paris (1875), vol. 2, p. 11, De l'interpretation de la nature, section iv.

[33] G. Gore, *The Art of Scientific Discovery: Or the General Conditions and Methods of Research in Physics and Chemistry*, pp. 15–16 and pp. 26–9, London, (1878).

[34] R. Feynman, *The Character of Physical Law*, p. 172, MIT Press, Cambridge, Mass. (1965).

[35] B. Glass, ref. 8. Note, however, that we cannot be sure that 'the universe is closed and finite' as the author asserts.

[36] In I. Good (ed.), *The Scientist Speculates*, p. 15, Basic Books, New York (1962).

[37] I first worked out this result during my D.Phil. oral examination in Oxford on 4 July 1977. The two examiners had both kept lists of typographical errors that they had found while reading the thesis and wondered how many had been missed. This argument supplied them with an answer. Fortunately, the prediction was not too large.

[38] A similar type of reasoning has been used, in a more limited domain, to estimate the number of new astronomical phenomena that remain unfound, in M. Harwit, *Cosmic Discovery*, MIT Press, Cambridge, Mass. (1981). Harwit thought that we should have found 90 per cent of all important classes of astronomical object by the year 2200.

[39] *Troilus and Cressida*, Act 3 scene 2.

[40] J.D. Barrow, *Theories of Everything*, Clarendon Press, Oxford (1991).

Chapter 4

[1] R. Trivers, Sociology and Politics, in E. White (ed.), *Sociobiology and Human Politics*, p. 33, Lexington Books, Lexington, Mass. (1981).

[2] This is sometimes called the 'Central Dogma' of evolutionary biology; see for example, E. Mayr, *One Long Argument*, Penguin Books, London (1991).

[3] S. Pinker, *The Language Instinct*, Penguin Books, London (1994).

[4] J. Seymour and D. Norwood, A game of life, *New Scientist*, **139**, (No. 1889), 23 (1993).

[5] R. Descharnes and G. Néret, *Dali*, Vols. 1 and 2, Taschen, Hohenzollernring (1994).

[6] S. Mithen, *The Prehistory of the Mind*, Thames and Hudson, London (1996).

[7] Towards the end of the nineteenth century, the Italian Giovanni Schiaparelli reported sightings of '*canali*' (channels) on Mars. In 1895 this inspired the American astronomer Percival Lowell to map many of these assumed Martian canals using a purpose-built telescope at Flagstaff, Arizona. More recently, the 1976 *Viking 1* Mars orbiter took photographs of the martian surface which show a peculiar rock formation, curiously resembling a human face, in relief; see D. Goldsmith, *The Hunt for Life on Mars*, p. 199, Dutton, New York (1997).

[8] G.C. Williams, *Plan and Purpose in Nature*, Orion, London (1997); R. Dawkins, *The Blind Watchmaker*, Norton, London (1986).

[9] L.F. Tóth, What the bees know and what they do not know, *Bulletin of the American Mathematical Society*, **70**, 468 (1964); S. Hildebrandt and A. Tromba, *Mathematics and Optimal Form*, W.H. Freeman, New York (1985).

[10] R. Rorty, 'Is the truth out there?', interview recorded in *The Times Higher Education Supplement*, 6 June, (1997), p. 18; for a fuller account, see R. Rorty, *Philosophy and the Mirror of Nature*, Blackwell, Oxford, (1980).

[11] N. Humphrey, *Soul Searching*, pp. 52–3, Chatto and Windus, London (1995).

[12] J.D. Barrow, *Pi in the Sky*, Oxford University Press (1992).

[13] English still retains some old words which describe small numbers of special things. For example, we talk of a brace of pheasants or a pair of shoes. This applies mostly to twos; when larger quantities are referred to, the special terms become rarer.

[14] Quoted in G. Stent, *The Coming of the Golden Age*, Natural History Press, New York (1969) p. 98.

[15] R. Voss and J. Clarke, $1/f$ (flicker) noise: a brief review. In *Proceedings of the 33rd Annual Symposium on Frequency Control*, pp. 40–6, Atlantic City (1975), and $1/f$ noise in music: music from $1/f$ noise. *Journal of the Acoustical Society of America*, **63**, 258 (1978); for further discussion see J.D. Barrow, *The Artful Universe*, Oxford Univrsity Press, (1995) ch. 5.

[16] L.B. Meyer, *Music, the Arts and Ideas*, University of Chicago Press, Chicago (1967). Meyer, it should be noted, propounded a theory of music which maintains that certain tonal sequences create definite emotional responses.

[17] see S. Mithen, op cit.

[18] Exodus 1, v. 7.

[19] I am grateful to John Casti for help locating this picture. It appears in J. Casti, *Five Golden Rules*, Wiley, NY (1996).

[20] D. Harel, *Algorithmics: The Spirit of Computing*, Addison Wesley, New York (1987).

[21] W. Rouse Ball, *Mathematical Recreations and Essays*, Macmillan, London (1982).

[22] The best ways of achieving this were found by P. Buneman and L. Levy, and also by T. Walsh, in 1980; see L. Levy, *Discrete Structures of Computer Science*, Wiley, New York (1980).

[23] Notice a technical point here; **NP** problems are defined in terms of the time needed to check the correctness of the solution. It should take much longer to find it (compare the time needed to solve a jigsaw puzzle with the time needed to check that it has been correctly completed).

[24] S.A. Cook, The complexity of theorem proving procedures, *Proceedings of the 3rd ACM Symposium on Theory of Computing*, p. 151, ACM, New York, (1971); M.R. Garey and D.S. Johnson, *Computers and Intractability: A Guide to the Theory of NP-Completeness*, Freeman, San Francisco (1979).

[25] J. Casti, The outer limits: in search of the 'unknowable' in science, *in* J. Casti and A. Karlqvist, *Boundaries and Barriers*, p. 27, Addison Wesley, New York (1996).

[26] A. Fraenkel, Complexity of protein folding, *Bulletin of Mathematical Biology*, **55**, 1199 (1993). For some further discussion which focuses on the subtle questions of the type of model of computation that should be adopted in these and other studies of complexity, see J. Traub, *in* J. Casti and A. Karlqvist, *Boundaries and Barriers*, p. 249, Addison Wesley, New York (1996).

[27] G. Rose, No Assembly Required, *The Sciences*, pp. 26–31 (Jan/Feb 1996).

[28] D.P. Di Vicenzo, Quantum computation, *Science*, **270**, 255 (1995); L.M. Adelman, Molecular computation of solutions to combinatorial problems, *Science*, **266**, 1021 (1994).

[29] M.R. Schroeder, *Number Theory in Science and Communication*, 2nd edn, p. 118, Springer, New York (1986).

[30] See *New Scientist*, No. 2080, p. 13, No. 2080.

[31] J. Diamond, *The Rise and Fall of the Third Chimpanzee*, pp. 204–5, Vintage, London.

[32] *The African Queen*, words spoken to Humphrey Bogart.

[33] Quoted in G. Dyson, *Darwin Among the Machines*, p. 108, Addison Wesley, New York, (1997).

Chapter 5

[1] *Die Fröhliche Wissenshaft*, (1886), IV

[2] Ch. 14, v. 28.

[3] W. Zurek, Thermodynamic cost of computation, algorithmic complexity and the information metric, *Nature*, **342**, 119 (1989).

[4] E. Mayr, *One Long Argument: Charles Darwin and the Genesis of Modern Evolutionary Thought*, Harvard University Press, New York (1991).

[5] F. Nietzsche, *The Will to Power*, in Collected Works, ed. O. Levy, new edn, T. N. Foulis, London (1964). For a modern approach to biological progress, see for example F.J. Ayala, Can 'progress' be defined as a biological concept?, *in* M. Nitecki (ed.), *Evolutionary Progress*, pp. 75–96, Chicago University Press (1988).

[6] See J.D. Barrow and F.J. Tipler, *The Anthropic Cosmological Principle*, Clarendon Press, Oxford. (1986). As yet, the discovery of a new form of carbon (carbon-60) has yet to have an impact on biochemical routes to the spontaneous evolution of life. Perhaps it will provide us with a new route to biological complexity?

[7] Approximately 25 per cent of the mass in the Universe is in the form of helium, of which about 1–2 per cent is made in the stars. The rest was produced when the Universe was about three minutes old. The remaining 75 per cent is almost all hydrogen.

[8] That is, with mass proportional to the volume (or the cube of the radius).

[9] Other creatures enter because of the structure of the food chain.

[10] A. Smith, *Essays on the Principles which lead and direct philosophical inquiries*, Ward, Lock & Co., London (1880).

[11] R. Feynman, *QED: The Strange Theory of Light and Matter*, Princeton University Press (1985).

[12] This accuracy is achieved by monitoring the behaviour of a binary pulsar; see C. Will, *Was Einstein Right?*, Basic Books, New York (1986); I. Ciufolini and J.A. Wheeler, *Gravitation and Inertia*, Princeton University Press (1995).

[13] It is possible that cosmological observations may require the presence of an additional force of Nature, sometimes referred to as the 'cosmological constant'. One can view this as an addition to the existing form of the law of gravitation, or as a new long-range force of Nature. In contrast to the inverse-square law of gravity, proposed by Newton, this force increases linearly with distance; since its effects are small, even over the scale of the Universe, they must be negligible in the solar system and on smaller scales.

[14] See J.D. Barrow, *The Artful Universe*. Oxford University Press (1995).

[15] Helium was identified as an element soon afterwards by Edward Frankland and Norman Lockyer. It was isolated in the laboratory by William Ramsey in 1895.

[16] N.S. Kardeshev, Transmission of information by non-terrestrial civilizations, *Soviet Astronomy*, **8**, 217 (1964); C. Sagan and I.S. Shklovskii, *Intelligent Life in the Universe*, p. 469, Dell, New York (1966), point out that Constantin Edwardovich Tsiolkovskii, a Russian space pioneer, speculated about astronomical manipulation of the environment in his 1895 book *Dreams of the Earth and Sky*. He claimed that we received only 5×10^{-10} of the total solar flux. We could use it all by colonizing the whole solar system, rebuilding asteroids into a chain of cities and controlling them and minor planets 'in the same way that we drive horses', using 'solar motors' (batteries), and this 'could easily support 3×10^{23} beings.'

[17] In 1960, Freeman Dyson considered the limits of energy and matter resources. At present, we can use less than 10^{-8} of the mass of the Earth. With a total energy consumption of 1–2 billion tons of coal per year we would expend an average of 3×10^{19} ergs per second and we would exhaust our fossil fuels in several hundred years. If average annual growth is merely 0.003 per cent, then in 1500 years energy requirements will be 3×10^{29} ergs per second, roughly 0.01 per cent of the luminosity of the Sun. Means of harvesting energy more deeply and widely could postpone our demise considerably, but the general problem remains.

[18] If the spectral index of power emitted by civilizations is steeper than 2.5, then the bright distant ones are more easily detected than the closest.

[19] There have even been suggestions that the bursts of gamma radiation, which astronomers have been trying to locate and explain for the past few years, are the hyperdrives from extraterrestrial spacecraft!

[20] J.D. Barrow and F.J. Tipler, *The Anthropic Cosmological Principle*, Clarendon Press, Oxford (1986).

[21] J.D. Barrow and F.J. Tipler, op. cit.

[22] A. Guth and S. Blau, *in* S.W. Hawking and W. Israel (eds), *300 Years of Gravitation*, Cambridge University Press (1987); E. Farhi and A. Guth, *Physics Letters*, **183** B, 149 (1987).

[23] A. Linde, Cosmology and Phase Transitions, *Reports on Progress in Physics*, **47**, 925 (1984).

[24] L. Smolin, Did the Universe Evolve?, *Classical and Quantum Gravity*, **9**, 173 (1992) and *The Life of the Cosmos*, Weidenfeld, London (1997).

[25] There are ambiguities, though, about what is meant by this rate. Also, it is quite likely that the assumed optima do not exist for certain constants of Nature.

[26] E.R. Harrison, The natural selection of universes containing intelligent life, *Quarterly Journal of the Royal Astronomical Society*, **36**, 193 (1995).

[27] Some of these constants appear to have origins in quasi-random events which could have fallen out differently; that is, they could have taken different values in universes governed by the same laws as our own. It is quite possible that all the constants might originate in a statistical way. Indeed, even something like the fact that we experience three dimensions of space might share this haphazard origin.

[28] F. Hoyle, *Religion and the Scientists*, SCM Press, London (1959). Hoyle's original prediction of carbon-12 energy levels is in *Astrophysical Journal Supplement*, **1**, 121 (1954).

[29] If the strong nuclear force were slightly stronger, then the dineutron and the diproton (that is, the helium-2 isotope) would exist as bound states and provide a very fast direct route for hydrogen burning in stars. This might well be a small change in the value of a constant which increases black hole production and so rules out Smolin's hypothesis, see note 24 above.

[30] F. Dyson, Energy in the Universe, *Scientific American*, **225**, 25 (Sept. 1971).

[31] This type of speculation began with J.D. Bernal's book *The World, the Flesh and the Devil*, (2nd edn, Indiana University Press (1969); see also Freeman Dyson's *Infinite in all Directions*, Basic Books, New York (1988).

[32] For the most extreme example see F.J. Tipler, *The Physics of Immortality*, Doubleday, New York (1994) and ref. 19.

[33] S.W. Hawking, *A Brief History of Time*, Bantam, New York (1988).

[34] W. Blake, *Songs of innocence*, (1789). G. Keynes (ed.), Oxford University Press (1970).

[35] Complexity need not exist only in spatial organization. It can also manifest itself in the arena of motion (what physicists call velocity space). Turbulent liquids, rushing from a tap or cascading down a waterfall, are prime examples; so are the intricate chaotic motions we see in the rings of Saturn. This raises the intriguing possibility that, if we define life as a manifestation of some critical level of complexity being achieved, then life may be possible in velocity space as well as in the more familiar three-dimensional position space that we inhabit.

[36] P. Bak, *How Nature Works: the science of self-organised complexity*, Springer, New York (1996).

[37] If the sand is made wet, or is ground finer, then a critical slope still arises but it will be different, as it will be if sugar, or rice, is substituted for sand.

[38] That is, the frequency of occurrence of events is proportional to a mathematical power of its size. This type of process is called self-similar.

[39] The probability of an avalanche of N sand grains must be proportional to N^{-a}, where a is some positive number.

[40] J. Deboer, B. Derrida, H. Flyvbjerg, A. Jackson, and T. Wettig, Simple model of self-organised biological evolution, *Physical Review Letters*, **73**, 906 (1994); K. Sneppen, P. Bak, H. Flyvbjerg, and M.H. Jensen, Evolution as a self-organized critical phenomenon, *Proceedings of the National Academy of Sciences of the USA*, **92**, 5209 (1995).

[41] P. Bak and C. Tang, Earthquakes as a self-organized critical phenomenon, *Journal of Geophysical Research* B **94**, 15, 635 (1989); A. Sornette and D. Sornette, Self-organized criticality and earthquakes, *Europhysics Letters*, **9**, 197 (1989).

[42] K. Nagel and M. Paczuski, Emergent traffic james, *Physical Review* E, **51**, 2909 (1995).

43 J.A. Scheinkman and M. Woodford, Self-organised criticality and economics fluctuations, *American Journal of Economics*, **84**, 417 (1994).

44 R. Voss and J. Clarke, 1/*f* noise in music and speech, *Nature*, **258**, 317 (1975), 1/*f* noise in music: music from 1/*f* noise, *Journal of the Acoustical Society of America*, **63**, 258 (1978). A detailed discussion of these results and some of their implications is given in Chapter 5 of my book *The Artful Universe*.

45 K.E. Drexler, *Engines of Creation*, p. 148, Fourth Estate, London (1990).

46 Note that this need not be true if the system is not closed. In that case the system can receive energy and information from outside. It is possible for its entropy to decrease locally at the expense of increases elsewhere. Life is a physical process that exploits this possibility far from equilibrium. A candle flame is another example.

47 J.C. Maxwell, *The Theory of Heat*, ch. 12, Longmans, Green, and Co., London (1871).

48 This name was given to it by Lord Kelvin, although Maxwell was not entirely happy about it.

49 L. Szilard, *Zeitschrift für Physik*, **53**, 840 (1929), reprinted in H.S. Leff and A.F. Rex, *Maxwell's Demon*, pp. 124–33, Princeton University Press (1990), as 'On the decrease of entropy in a thermodynamic system by the intervention of intelligent beings'. If one ignores the second law then it is possible to calculate how efficient the demon can be as a source of power. The only limit used is the Heisenberg Uncertainty Principle. It is found that for a dilute gas at room temperature, with a volume equal to that of a typical room, a temperature difference of two degrees between two halves of the room would take more than 1000 years to achieve; see H.S. Leff, Thermal efficiency at maximum work output: new results for old heat engines, *American Journal of Physics*, **55**, 602 (1987) and Maxwell's demon, power and time, *American Journal of Physics*, **58**, 135 (1990).

50 Notice the inclusion of the requirement that the slate be wiped clean in preparation for the next measurement. The essential nature of this part of the process was identified by Rolf Landauer only in 1961; see H. Leff and A. Rex, ref. 49, for a list of papers.

51 J.D. Bekenstein, Energy cost of information transfer, *Physical Review Letters*, **46**, 623 (1981).

52 E. Wigner, Relativistic invariance and quantum phenomena, *Reviews of Modern Physics*, **29**, 255 (1957); J.D. Barrow, Wigner inequalities for a black hole, *Physical Review* D **54**, 6563 (1996).

53 P.D. Pesic, The smallest clock, *European Journal of Physics*, **14**, 90 (1993).

54 D.T. Spreng, On time, information, and energy conservation, ORAU/IEA–78–22(R). Institute for Energy Analysis, Oak Ridge Assoc. Universities, Oak Ridge. Tennessee (Dec. 1978).

55 A.M. Weinberg, On the relation between information and energy systems: a family of Maxwell's demons, Maxwell's Demon, text of lecture delivered 27 Oct. 1980 to the National Conference of the Association of Computing Machinery, Nashville, reproduced in Leff and Rex, ref. 49, p. 116.

56 K. Clark, *Civilisation*, p. 345, John Murray, London (1971).

57 J. Maynard Smith, *Evolutionary Genetics* p. 125, Oxford University Press (1989).

58 A.C. Clarke, *Childhood's End*, Sidgewick and Jackson, London (1954).

59 G. Marx (ed.), *Bioastronomy—the next steps*, Kluwer, Dordrecht (1988).

60 E. Regis (ed), *Extraterrestrials: science and alien intelligence*, Cambridge University Press, (1985).

[61] M.D. Papagiannis, *Quarterly Journal of the Royal Astronomical Society*, 25, 309 (1984).

[62] See the discussion by M. Ridley, *The Origins of Virtue*; W. Irons, How did morality evolve?, *Zygon*, 26, 49 (1991); F.J. Ayala, The difference of being human: ethical behaviour as an evolutionary byproduct, in *Biology, Ethics, and the Origins of Life*, H. Rolston (ed.), Jones and Bartlett, Boston (1995); F. de Waal, *Good Natured: The Origins of Right and Wrong in Humans and Other Animals*, Harvard University Press, Cambridge, Mass. (1995); P. Hefner, Theological perspectives on morality and human evolution, *in* M. Richardson and W.J. Wildman (eds), *Religion and Science: History, Method, Dialogue*, Routledge, New York (1996).

[63] H. Moravec, *Mind Children: The Future of Robot and Human Intelligence*, Harvard University Press, Cambridge, Mass. (1988).

[64] O. Stapleton, *Star Maker*, Dover, New York (1968).

[65] A. Rice, *The Witching Hour*, Knopf, New York, (1990).

[66] A. C. Clarke, *Superiority*, in C. Fadiman (ed.), *Fantasia Mathematica*, pp. 110–20, Simon and Schuster, New York (1958). The story was first published in 1951.

[67] J.D. Barrow, *The Origin of the Universe*, Weidenfeld, London (1994).

[68] From M.F. Crommie, C.P. Lutz, D.M. Eigler (1993), *Science*, 262, 218; *Phys. Rev.* B48, 2851; *Nature*, 363, 524.

[69] Photograph courtesy of Franco Nori; see M. Bretz, J. Cunningham, P. Kurczynski, and F. Nori, Imaging of avalanches in granular materials, *Physical Review Letters*, 69, 2431 (1992); and *Science News*, 142, 231 (1992).

Chapter 6

[1] R. Estling, *The Skeptical Inquirer*, Spring issue, (1993).

[2] D. Adams, *Mostly Harmless*, p. 1, Heinemann, London (1992).

[3] S. Goodwin, *Hubble's Universe*, Anchor, London, (1997).

[4] S.W. Weinberg, *The First Three Minutes*, Basic Books, New York (1975); J.D. Barrow and J. Silk, *The Left Hand of Creation*, Basic Books, New York (1983) and 2nd edn Oxford University Press, New York and Penguin Books, London (1995); J.D. Barrow, *The Origin of the Universe*, Weidenfeld, London (1994).

[5] The word 'theory' has come to have a slightly negative connotation in ordinary parlance, conveying the idea that something is wildly speculative, uncertain, or hare-brained. In science it is a term given to any system of ideas or mathematical equations. These are all provisional in the sense that experiment may one day falsify them. However, some theories, like Einstein's general theory of relativity, have made astonishingly successful predictions. When good theories are superseded, what generally happens is that they turn out to be a limiting case of a more general description, just as Newton's theories are the limiting cases of Einstein's when velocities are far smaller than that of light and gravitational fields are weak.

[6] Matthew ch. 13, v. 12.

[7] A good account which provides a careful distinction of some of these uncertainties is E.R. Harrison, *Cosmology*, Cambridge University Press (1981), although it is now rather out of date since many significant developments in observational and theoretical cosmology, especially from particle physics, occurred after it was written.

[8] Its size is given by the speed of light multiplied by the time that the expansion has been going on. At present this is roughly $(3 \times 10^{10} \text{ cm/sec}) \times (10^{17} \text{ sec}) \approx 10^{27} \text{ cm}$. At any

time t (in seconds) after the expansion has begun (at $t = 0$, say), the mass within this horizon sphere is about $10^5 \times t$ times the mass of our Sun. Conversely, as we go back to the 'beginning' (as t approaches zero) the mass of matter and radiation within our horizon goes to zero.

9 Remarkably, Newton's theory of gravitation does make this distinction. In Newton's theory universes with infinite volume cannot exist; see F.J. Tipler, Newtonian Cosmology revisited, Monthly Notices of the Royal Astronomical Society **282**, 206–10 (1996).

10 From *The Life of William Thomson, Baron Kelvin of Largs*, Macmillan, London (1910).

11 This is because the Universe contains irregularities.

12 The acceleration is so rapid that only a very brief period is required to do this.

13 G. Smoot and K. Davidson, *Wrinkles in Time*, Morrow, New York (1994).

14 C. Misner, Transport processes in the primeval fireball, *Nature*, **214**, 40 (1967).

15 J.D. Barrow and R. Matzner, The homogeneity and isotropy of the Universe, *Monthly Notices of the Royal Astronomical Society*, **181**, 719–28, (1977).

16 This possibility exists if the scalar field has many possible final resting places, rather than just one. That is, it is like throwing a ball on to a piece of corrugated sheeting, with its many undulations, rather than into a bowl with a resting place at the bottom.

17 Remark to William Allingham, quoted in D.A. Wilson and D. Wilson McArthur, *Carlyle: Carlyle in Old Age*, vol. 6, Kegan Paul, London, (1934).

18 It should be stressed that this time interval is chosen for illustration's sake only. We cannot detect global changes in the visible universe over short intervals of time. Significant changes to the expansion of the visible universe now occur only over periods of billions of years.

19 A. Linde, The inflationary universe, *Physics Today*, **40**, (9) 61 (1987).

20 A detailed discussion of a number of the requirements can be found in J.D. Barrow and F.J. Tipler, *The Anthropic Cosmological Principle*, Clarendon Press, Oxford (1986).

21 J.D. Barrow and A. Liddle, Can inflation be falsified?, *General Relativity and Gravitation Journal*, **29**, 1501–8 (1997).

22 An immense amount of data will be provided by the MAP and Planck Surveyor missions. This may enable us to determine the properties of the scalar field that inflated our region of the Universe. We might then be able to say whether its properties were consistent with the eternal inflationary sequence, but we could not confidently rule it out because we expect there to be a huge number of fields of this type. Elsewhere, other fields may be doing the inflating of other bubbles.

23 F. Hoyle, A new model for the expanding Universe, *Monthly Notices of the Royal Astronomical Society*, **108**, 372 (1948); H. Bondi and T. Gold, The steady-state theory of the expanding Universe, *Monthly Notices of the Royal Astronomical Society*, **108**, 252 (1948).

24 F. Hoyle and J. Narlikar, *Proceedings of the Royal Society* A, **290**, 143, 162 (1966).

25 This could be accommodated by introducing a 'scale' over which the steadiness is to be found.

26 Quoted in R. Lewin, Why is development so illogical?, *Science*, **224**, 1327 (1984).

27 This is particularly evident in the writings of S.J. Gould and R. Dawkins. The opposing view is put forward in considerable detail by S. Kauffman, *The Origins of Order*, Oxford University Press, New York (1993), and *At Home in the Universe*, Oxford University Press, New York (1995).

28 A. Linde, The self-reproducing inflationary universe, *Scientific American* 5 (May), 32

(1994). A nice popular account of 'other' universes can be found in M.J. Rees, *Before the Beginning*, Simon and Schuster, New York, (1997).

29 J. Maynard Smith, *Evolutionary Genetics*, Oxford University Press (1989).

30 Attempts have been made to discover whether it is more or less likely for a universe to have a natural topology if it has a quantum origin. Unfortunately, the answer is not clear-cut: different ways of doing the calculation give different answers at present.

31 J. Levin, J.D. Barrow, E. Bunn, and J. Silk, Flat spots: topological signatures of an open universe in *COBE* sky maps, *Physical Review Letters*, **79**, 974–8 (1997).

32 This works both ways. Some cosmologists favoured the Big Bang theory in the 1950s because it resembled the Judaeo-Christian account of creation a finite time ago, whereas some of the proponents of the steady-state theory said they were attracted to it because was so unlike this traditional cosmology.

33 R. Penrose, *The Emperor's New Mind*, Oxford University Press (1989). In particular, it is argued that there must exist 'laws' governing the structure of singularities that dictate that the initial state of the Universe is overwhelmingly likely to be highly ordered. The final state will be extremely disordered, in accord with the growth of entropy, governed by the second law of thermodynamics.

34 See J.D. Barrow, *The Origin of the Universe*, Orion, London (1994), ch. 2, for a discussion.

35 It could increase like an exponential of the time, for example. There must be a smallest possible change in the entropy in order for a minimum entropy value to exist.

36 For a more detailed non-technical account, see J.D. Barrow and J. Silk, *The Left Hand of Creation*, 2nd. edn, Penguin Books, London (1994).

37 R. Penrose, Gravitational collapse and space-time singularities, *Physical Review Letters*, **14**, 57 (1965).

38 S.W. Hawking and R. Penrose, The Singularities of Gravitational Collapse and Cosmology, *Proceedings of the Royal Society*, **A 314**, 529 (1970).

39 A. Guth, The Inflationary Universe, *Physical Review*, **D 23**, 347 (1981) and *The Inflationary Universe*, Addison Wesley, New York (1997).

40 J.P. Luminet, *Black Holes*, Cambridge University Press (1987).

41 S.W. Hawking, Black hole explosions, *Nature*, **248**, 30 (1974).

42 D. Adams, *Mostly Harmless*, p. 25, W. Heinemann, London, (1992).

43 J.D. Barrow, Observational limits on the time-evolution of extra spatial dimensions, *Physical Review* D **35**, 1805 (1987).

44 J.D. Prestage, R.L. Tjoelker, and L. Malecki, *Physical Review Letters*, **74**, 3511 (1995).

45 M. Drinkwater, J. Webb, J. D. Barrow and V. V. Flambaum, New Limits on the variation of physical constants, *Monthly Notices of the Royal Astronomical Society* (in press, 1997).

46 M. Maurette, The Oklo Reactor, *Annual Reviews of Nuclear and Particle Science*, **26**, 319 (1976).

47 A.I. Shylakhter, *Nature*, **264**, 340 (1976); F. Dyson and T. Damour, The Oklo Bound on the time variation of the fine-structure constant revisited, *Nuclear Physics*, **B 480**, 37 (1997).

Chapter 7

1 J.D. Barrow, *Pi in the Sky: counting, thinking, and being*, Oxford University Press (1992).

[2] R. Rosen, *Life Itself*, Columbia University Press (1991).

[3] A. Rice, *The Witching Hour*, Knopf, New York (1990).

[4] A nice collection is contained in P. Hughes ad G. Brecht, *Vicious Circles and Infinity: an anthology of paradoxes*, Penguin Books, London (1978).

[5] One of the oldest logical paradoxes which Tarski's analysis unravels is the 'paradox of the liar', attributed to the sixth-century BC Greek philosopher, Eubulides of Megara. His original version of the paradox required the liar to answer the question 'Do you lie when you say that you are lying?' If the liar says 'I am lying', then clearly he is not lying, because if a liar says he is a liar when he really is a liar, then he is speaking the truth. But if the liar says 'I am not lying', then it is true that he is lying, and, consequently, he is lying.

[6] See J.D. Barrow, *Pi in the Sky*, Oxford University Press (1992), p. 18.

[7] R. Smullyan, *This Book Needs No Title*, p. 139, Prentice-Hall, New York (1980).

[8] For example, in the dipolar opposites of process theologians like the philosopher Charles Hartshorne, *A Natural Theology for our Time*, Open Court, Law Salle (1967).

[9] In *De Omnipotentia Dei*, see p. 179, P. Nahin, *Time Machines*, American Institute of Physics, New York (1993). St. Damian appears as Don Pedro Damián in the short story 'The Other Death' by Jorge Luis Borges (see *The Aleph*, ed. N.T. di Giovanni, Dutton, New York, 1978, p. 103). The narrator pieces together a strange sequence of events after discovering the theological case for backwards causality in the treatise *De Omnipotentia*.

[10] C.S. Lewis, *Miracles*, Collins, London (1947).

[11] C.J. Isham and J.C. Polkinghorne, The debate over the Block Universe, in R.J. Russell, N. Murphy, and C.J. Isham (eds), *Quantum Cosmology and The Laws of Nature*, pp. 135–44, Vatican Observatory, Vatican City (1993).

[12] P. Nahin, *Time Machines*, p. 41, American Institute of Physics, New York (1993).

[13] M. Kaku, *Hyperspace*, Oxford University Press, New York (1994).

[14] K. Gödel, An example of a new type of cosmological solution of Einstein's field equations of general relativity, *Reviews of Modern Physics*, 21, 447 (1949).

[15] Einstein's equations describe the properties of an infinite number of different possible universes obeying the same law of gravitation, of which our universe is just one example, specified by its starting state.

[16] H. Weyl, *Space, Time, and Matter*, trans. H. Brose, Methuen, London (1922).

[17] D. Piper, *Observatory Magazine*, 97, 10P (Oct. 1977).

[18] S.W. Hawking, The chronology protection hypothesis, *Physical Review* D, 46, 603 (1992); M. Visser, *Lorentzian Wormholes-from Einstein to Hawking*, American Institute of Physics, New York (1995).

[19] R. Silverberg, *Up the Line*, Ballantine, New York (1969).

[20] Nahin, op. cit., p. 167.

[21] See J.D. Barrow and F.J. Tipler, *The Anthropic Cosmological Principle*, ch. 9, Clarendon Press, Oxford, (1986).

[22] M.R. Reinganum, Is time travel possible?: A financial proof, *Journal of Portfolio Management* 13, 10–12 (1986).

[23] K. Gödel, A remark about the relationship between relativity theory and idealistic philosophy. In *Albert Einstein: Philosopher–Scientist*, Vol. 7 of The Library of Living Philosophers, ed. P.A. Schilpp, Open Court, Evanston IL (1949).

[24] D.B. Malament, Time travel in the Gödel Universe, *Proceedings of the Philosophy of Science Association*, 2, 91–100 (1984).

[25] D. Lewis, The paradoxes of time travel, *American Philosophical Quarterly*, 13, 15–152 (1976).

[26] L. Dwyer, Time travel and changing the past, *Philosophical Studies*, 27, 341–50 (1975); see also Time travel and some alleged logical asymmetries between past and future, *Canadian Journal of Philosophy* 8, 15–38 (1978); How to affect, but not change, the past, *Southern Journal of Philosophy*, 15, 383–5 (1977).

[27] D. Deutsch, Quantum mechanics near closed timelike lines, *Physical Review* D, 44, 3197 (1991).

[28] See D. Deutsch, *Physical Review* D, 44, 3197 (1991) and D. Deutsch and M. Lockwood, *Scientific American* (March 1994) 270, 68–74; and D. Deutsch, *The Fabric of Reality*, Penguin Books, London (1997).

[29] See J.D. Barrow, *Pi in the Sky*, Oxford University Press (1992) for further details.

[30] For a summary of the mathematical results, with proofs, see G. Birkhoff and S. Maclane, *A Survey of Modern Algebra*, Macmillan, New York (1964).

[31] E.T. Bell, *Men of Maths*, vol. 1, p. 311, Schuster, New York (1965); S.G. Shanker (ed.) *Gödel's Theorem in Focus*, p. 166, Routledge, London (1988).

[32] The requirement that statements make logical sense means that not every integer is a Gödel number.

[33] At the start of this chapter we looked at paradoxical statements which asserted their own falsity ('This statement is false'). The logician Leon Henkin has given his name to statements which are automatically true ('This sentence is provable'). Henkin sentences are self-proving sentences. The proof was given by Löb, who proved the following interesting theorem relating statements and metastatements. If we have a system, S, in which the statement '*if A is provable in S then A is true*' is provable in S, then A is provable in S. This theorem of Löb's implies one of Gödel's incompleteness theorems as a particular example, if we take the formula $0 = 1$ to be the statement labelled A. We can then conclude that the consistency of S is not provable in S. For further discussion about incompleteness theorems, see C. Smorynski, The Incompleteness Theorems, *in* J. Barwise (ed.), *Handbook of Mathematical Logic*, North Holland, (1977). For a serious popularization, see R. Smullyan, *Forever Undecided; A Puzzle Guide to Gödel*, Oxford University Press (1987).

[34] J. Myhill, Some philosophical implications of mathematical logic, *Review of Metaphysics*, 6, 165 (1952). For further discussion see J.D. Barrow, *Theories of Everything*, p. 209, Clarendon Press, Oxford (1991), and D. Hofstadter, *Metamagical Themas*, p. 539, Basic Books, New York (1985).

[35] In logic, they are called 'recursive', 'renotrec' (= recursively enumerable but not recursive), and 'productive', respectively.

[36] This is the last theorem in Conway's book *On Numbers and Games*, p. 224, Academic Press, New York (1976).

Chapter 8

[1] Quoted in P. J. Davis and D. Park (eds), *No Way—Essays on the Nature of the Impossible*, p. 98, W.H. Freeman, New York (1987), Yarmolinsky served in the administrations of US Presidents Kennedy, Johnson, and Carter.

[2] R. Rucker, *Infinity and the Mind*, p. 165, Harvester, Sussex (1982).

[3] S. Jaki, *Cosmos and Creator*, p. 49, Scottish Academic Press, Edinburgh (1980).

[4] S. Jaki, *The Relevance of Physics*, p. 129, Chicago University Press (1966).

[5] G. Chaitin, *Information, Randomness and Incompleteness*, World Scientific, Singapore (1987).

[6] S. Lloyd, The calculus of intricacy, *The Sciences*, 38–44, (Sept/Oct 1990); J.D. Barrow, *Pi in the Sky*, pp. 139–40, Oxford University Press (1992).

[7] See our discussion of the Maxwell demon in Chapter 5.

[8] K. Gödel, What is Cantor's Continuum Problem?, *Philosophy of Mathematics*, ed. P. Benacerraf and H. Putnam, Prentice-Hall, Englewood Cliffs, NJ (1964), p. 483.

[9] Recorded by Chaitin, *see* J. Bernstein, *Quantum Profiles*, Basic Books, New York (1991), pp. 140–1, and K. Svozil, *Randomness and Undecidability in Physics*, p. 112, World Scientific, Singapore (1993).

[10] S. Wolfram, *Cellular Automata and Complexity, Collected Papers*, Addison Wesley, Reading Mass. (1994).

[11] C. Calude, *Information and Randomness—An Algorithmic Perspective*, Springer, Berlin (1994); C. Calude, H. Jürgensen, and M. Zimand, Is independence an exception?, *Applied Mathematics and Computing* 66, 63 (1994); K. Svozil, in J.L. Casti and A. Karlqvist (eds), *Boundaries and Barriers: on the limits of scientific knowledge*, p. 215, Addison Wesley, New York (1996).

[12] The decision procedure is in general double-exponentially long, though; that is, the computational time required to carry out N operations grows as $(2^N)^N$. Presburger arithmetic allows us to talk about positive integers and about variables whose values are positive integers. If we enlarge it by permitting the concept of sets of integers to be used, then the situation becomes almost unimaginably intractable. It has been shown that this system does not admit even a K-fold exponential algorithm, for any finite K. The decision problem is said to be non-elementary in such situations: the intractability is unlimited.

[13] That the terms in the sum get progressively smaller is a necessary but not a sufficient condition for an infinite sum to be finite. For example, the sum $1+1/2+1/3+1/4+1/5+\ldots$ is infinite in value. It does not converge to a finite limit

[14] R. Rosen, On the limitations of scientific knowledge, in J.L. Casti and A. Karlqvist (eds), *Boundaries and Barriers: on the limits of scientific knowledge*, p. 199, Addison Wesley, New York (1996).

[15] John A. Wheeler has speculated about the ultimate structure of space-time being a form of 'pregeometry' obeying a calculus of propositions restricted by Gödel incompleteness. We are proposing that this pregeometry might be simple enough to be complete; see C. Misner, K. Thorne, and J.A. Wheeler, *Gravitation*, pp. 1211–2, W.H. Freeman, San Francisco (1973).

[16] The situation in superstring theory is still very fluid. There appear to exist many different, logically self-consistent superstring theories, but there are strong indications that they may be different representations of a much smaller number (perhaps even just one) theory that is called M-theory.

[17] N.C. da Costa and F. Doria, *International Journal of Theoretical Physics*, 30, 1041 (1991); *Foundations of Physics Letters*, 4, 363 (1991).

[18] Actually, there are other more complicated possibilities clustered around the dividing line between these two simple possibilities, and it is these that provide the indeterminacy of the problem in general.

[19] R. Geroch and J. Hartle, Computability and physical theories, *Foundations of Physics*,

16, 533 (1986). The problem is that the calculation of a wave function for a cosmological quantity involves the sum of quantities evaluated on every four-dimensional compact manifold in turn. The listing of this collection of manifolds is uncomputable.

[20] M.B. Pour-El and I. Richards, A computable ordinary differential equation which possesses no computable solution, *Annals of Mathematical Logic*, **17**, 61 (1979); The wave equation with computable initial data such that its unique solution is not computable, *Advances in Mathematics*, **39**, 215 (1981); Non-computability in models of physical phenomena, *International Journal of Theoretical Physics*, **21**, 553 (1982).

[21] J.F. Traub and A.G. Werschulz, Linear ill-posed problems are all solvable on the average for all gaussian measures, *The Mathematical Intelligencer*, **16**, (2), 42 (1994).

[22] S. Wolfram, Undecidability and intractability in theoretical physics, *Physical Review Letters*, **54**, 735 (1985); Origins of randomness in physical systems, *Physical Review Letters*, **55**, 449 (1985); Physics and computation, *International Journal of Theoretical Physics*, **21**, 165 (1982); J.F. Traub, Non-computability and intractability: does it matter to physics? (1997).

[23] If the metric functions are polynomials, then the problem is decidable, but it is computationally double-exponential. If the metric functions are allowed to be sufficiently smooth, then the problem becomes undecidable; see the article on algebraic simplification by Buchberger and Loos in Buchberger, Loos, and Collins, *Computer Algebra: Symbolic and Algebraic Computation*, 2nd edn, Springer, Vienna (1983). I am grateful to Malcolm MacCallum for supplying these details.

[24] S. Jaki, *The Relevance of Physics*, p. 129.

[25] E. Nagel and J. Newman, *Gödel's Proof*, p. 100, Routledge, London (1959), and *Scientific American*, **194**, 71 (June 1956).

[26] M. Scriven, in *The Compleat Robot: A Guide to Androidology*, ed. S. Hook, Collier Books, New York (1961).

[27] J. Lucas, Minds, Machines, and Gödel, *Philosophy*, **36**, 120 (1961).

[28] D. Hofstadter, *Metamagical Themas*, pp. 536–7 and bibliography, Basic Books, New York (1985).

[29] R. Penrose, *The Emperor's New Mind*, Oxford University Press (1989) and *Shadows of the Mind*, Oxford University Press (1994).

[30] R. Rucker, *Infinity and the Mind*, pp. 162–3, Harvester, Sussex (1982).

[31] The entire issue of the journal *Behavioural and Brain Sciences*, vol. 13, issue 4 (1990) was devoted to this debate.

[32] A. Kenny, H.C. Longuet-Higgins, J.R. Lucas, and C.H. Waddington, *The Nature of the Mind: Gifford Lectures 1971–1973*, Edinburgh University Press (1972), pp. 152–4.

[33] J. McCarthy, Review of *The Emperor's New Mind*, *Bulletin of the American Mathematical Society*, **23**, 606 (1990). McCarthy is the inventor of the LISP computer language.

[34] I. Singer, quoted in *The Times* (London) 'Diary', 21 June 1982.

[35] K.R. Popper, *British Journal for the Philosophy of Science*, **1**, 117, 173 (1950).

[36] K. Svozil, Undecidability everywhere, in *Boundaries and Barriers: on the limits of scientific knowledge*, ed. J.L. Casti and A. Karlqvist, Addison Wesley, New York (1996). Popper uses the fable of Achilles and the Tortoise to make the same point.

[37] D. Mackay, *The Clockwork Image*, p. 110, Inter-Varsity Press, London (1974).

[38] op. cit., p. 110.

[39] op. cit., p. 82.

[40] Science is difficult because there is no automatic way of finding the patterns which link together different properties of the real world. Remarkably, this suspicion can be expressed as a theorem, proved by E.M. Gold in 1967 (*Information and Control*, **10**, 447). It shows that if we have some mechanistic intelligence, like a computer or some other form of human or artificial intelligence, then there is a rule linking its input and output data which it cannot discover. That is, there is no systematic recipe for finding rules which link its input to its output. An interesting example of a finite automaton which possesses a property of complementarity, of the sort possessed by quantum mechanics, is given by K. Svozil in chapter 10 of this book *Randomness and Undecidability in Physics*, World Scientific, Singapore (1993). This book is the most important resource for the study of the interface between mathematical undecidability and physics. See also J.F. Traub, Do negative results from formal systems limit scientific knowledge?, *Complexity*, **3** (1), 29 (1997).

[41] *Le Rhinocéros* (1959), act 3.

[42] Simon's argument even appears in textbook treatments of political decision theory, like S. Brams's, otherwise excellent, *Paradoxes in Politics*, pp. 70–7, Free Press, New York (1976).

[43] K.E. Aubert and H.A. Simon, *Social Science Information* **21**, 610 (1982).

[44] H. Simon, Bandwagon and underdog effects in election predictions, *Public Opinion Quarterly*, **18**, 245 (Fall issue, 1954). The title refers to two contrasting reactions that election forecasting might have on voters' preferences. The 'bandwagon effect' is the tendency of some voters to vote with the candidate who is ahead in the opinion polls; the 'underdog effect' is the opposite tendency, of some voters to vote (sympathetically?) for the candidate who is behind in the polls.

[45] See any textbook on calculus for an explanation of this theorem; for example, the book *What is mathematics?* by R. Courant, H. Robbins, and I. Stewart, and in this specific context, K.E. Aubert, Accurate prediction and fixed point theorems, *Social Science Information*, **21**, 323 (1982).

[46] K. Aubert, Spurious mathematical modelling, *The Mathematical Intelligencer*, **6**, 59 (1984).

[47] D.H. Wolpert, An incompleteness theorem for calculating the future *Santa Fe Inst. preprint* (1996).

[48] It should be pointed out that not all mathematicians accepted Hilbert's dictum. Luitzen Brouwer argued that only entities which could be constructed in a finite number of separate deductive steps from the 'intuitively' given natural numbers should be said to exist within mathematics. This 'intuitionism' outlaws all sorts of concepts and forms of argument which mathematicians were in the habit of using. It never attracted widespread support. Hilbert condemned it as having the same effect on the mathematician as 'prohibiting the boxer the use of his fists'.

[49] J.D. Barrow, *Pi in the Sky: counting, thinking and being*, pp. 284–92, Oxford University Press (1992).

[50] Gödel wrote that mathematical objects like sets seem 'to me that the assumption of such objects is quite as legitimate as the assumption of physical bodies and there is quite as much reason to believe in their existence', in P. Schilpp (ed.), *The Philosophy of Bertrand Russell*, p. 137, Evanston, Chicago (1944).

[51] F.P. Adams, *Nods and Becks*, p. 206, (1944).

[52] I can remember being a member of a committee meeting to decide on the

appointment of one of three candidates to a job that took just such a vote at one stage in its deliberations.

53 A. Taylor, *Mathematics and Politics: Strategy, Voting, Power and Proof*, Springer, New York (1995).

54 A minor error in the original proof of the theorem was corrected in the second edition of Arrow's book *Social Choice and Individual Values*, Yale University Press (1963). The first edition was published in 1951.

55 C.E. Green, *The Decline and Fall of Science*, Hamilton, London, (1976).

56 This work is reviewed in D. Black, *Theory of Committees and Elections*, Cambridge University Press (1958).

57 A. Sen, A possibility theorem on majority decisions, *Econometrica*, 34, 491 (1966) and *Collective Choice and Social Welfare*, Holden-Day, San Francisco (1970).

58 S. Brams, *Paradoxes in Politics*, p. 142, Free Press, New York (1976).

59 F.J. Tipler, *The Physics of Immortality*, Doubleday, New York (1994).

60 See for example ref. 50.

61 Adapted from D. Harel, *Algorithmics: the spirit of computing*. Addison Wesley, New York (1987).

Index

DATE